PRACTICAL APPROACH TO PATTERN CLASSIFICATION

For my wife and parents

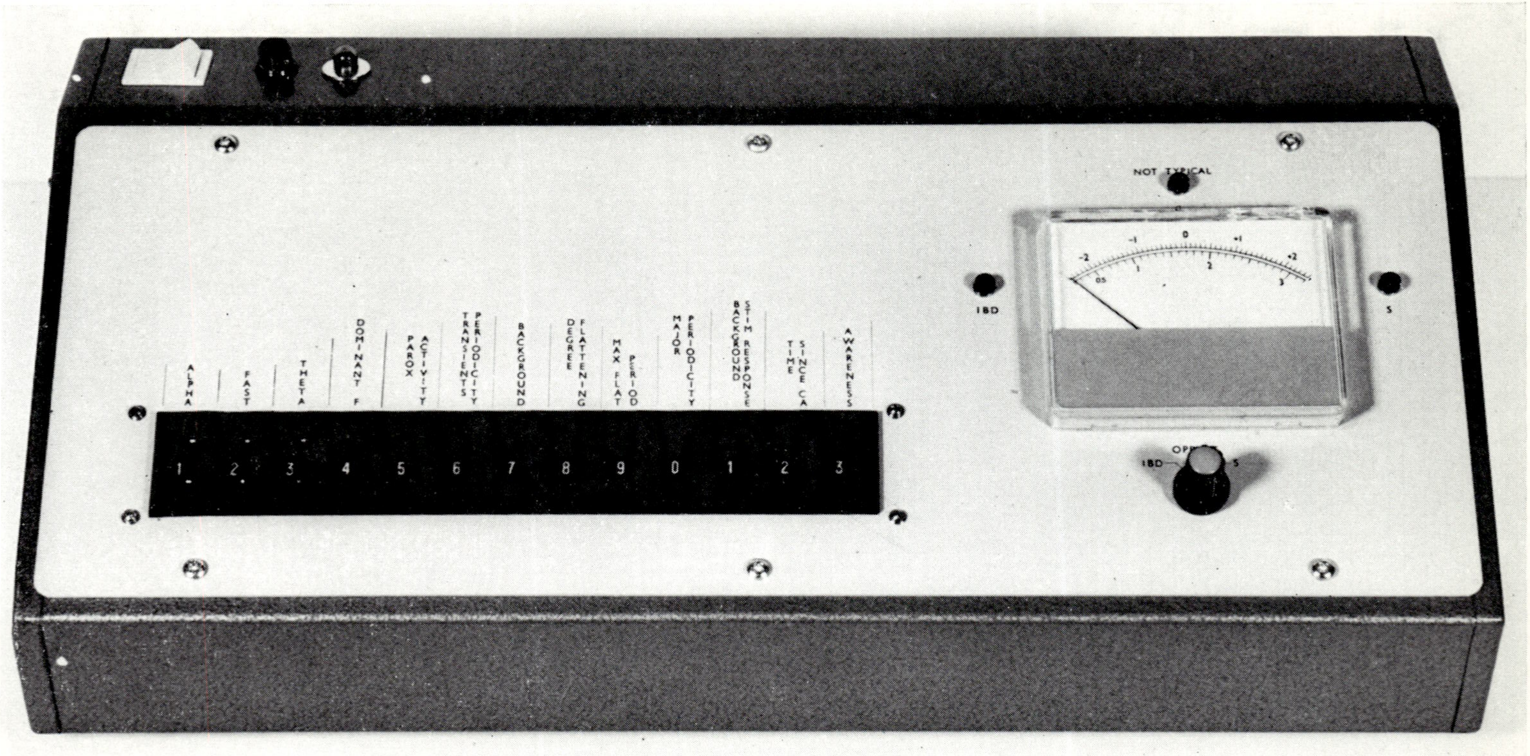

A pattern classifier built by D. E. Maynard, The London Hospital, to predict the outcome of cerebral anoxia caused by cardiac arrest.

PRACTICAL APPROACH TO PATTERN CLASSIFICATION

Bruce G. Batchelor

The University of Southampton
Southampton, England

PLENUM PRESS • LONDON AND NEW YORK

Library of Congress Catalog Card Number: 74-1614
ISBN 0-306-30796-0

Plenum Publishing Company Ltd
4a Lower John Streeet
London W1R 3PD
Telephone 01-437 1408

U.S. Edition published by
Plenum Publishing Corporation
227 West 17th Street
New York, New York 10011

Printed in Northern Ireland at The Universities Press, Belfast

Preface

Pattern classification is part of a wider subject, Pattern Recognition, which is concerned with the construction of machines that can perform such tasks as reading, recognising speech or diagnosing disease. The common element in these problems is the need for a device that can make decisions. The study of such machines is called pattern classification.

Pattern Recognition is still a dream, but what a dream! Suppose that a consensus opinion of the most experienced medical practitioners could be 'captured' in a machine that can emulate (some of) their diagnostic skills. Such a machine could be mass-produced and used all over the world. That particular dream is a long way off, but a little of it already exists. I hope that, by writing this book, I may be able to make a bit more come true, a little sooner. I am convinced that with the enthusiastic co-operation of specialists such as doctors, engineers and chemists we could solve many of the problems outlined in Chapters 2 and 7. Pattern recognition research is almost always expensive, although in many applications its aim is to save money in the future. I hope that this book will help to make more people aware of the possibilities and needs.

Our approach is heuristic, pragmatic and experimental, whereas most of the previous books on this subject rely heavily on mathematics and place little emphasis on the engineering aspects. These engineering problems are, at least, as important as the development of effective algorithms. Indeed, we already possess a procedure, which guarantees a 'solution' to the pattern classification problem. Unfortunately, this algorithm requires the use of an infinite store. This type of solution is of no value to an engineer; we must take into account the cost and speed of the procedure. Tukey expressed a similar idea in connection with statistical tests; he argued that *practical power* is the most meaningful measure of the usefulness of a test. He realized that *practical power* is the product of *mathematical power* and the *probability* that the procedure will be used. This probability is a function of the speed of the test and the cost of its implementation. The same argument can be applied to pattern classification procedures. Occam's Razor succinctly expresses another important principle which will guide us, namely that: *it is vain to do with more what can be done with less.* Furthermore, we should not feel constrained to use only those techniques which have received the benefit of detailed mathematical analysis. We can (and shall) propose procedures, based solely on our intuition, and evaluate them by experiment. We should not wait for a full mathematical analysis, before we apply these procedures. The pragmatic, experimental approach has a significant rôle in achieving faster progress in those situations where mathematical theory is poorly developed. Hopefully, the mathematical respectability of our techniques will follow, but we cannot wait for it, because there are many urgent problems awaiting the application of some really effective classification procedures.

Wherever possible, I have specified the procedures exactly, although there are still some areas where the intelligence of an experienced data analyst is needed to design a classifier successfully. I have tried to indicate the criteria that the user should apply in these situations. At Southampton, and elsewhere, research workers are still trying to eliminate the need for a skilled operator to 'drive' the procedures. At the moment we can only pass on our experience and offer to assist in applying the procedures to practical situations.

Bruce G. Batchelor
Southampton
April 1974

Contents

Chapter 8

Appendix I

Appendix 2

Chapter 1

Intelligent Decisions by Machine

1.1 DECISIONS

If the reader were asked to list those qualities that he regarded as being characteristic of intelligence he would almost certainly include the ability to analyse complex situations and make decisions about them. Normal everyday life requires so many decisions that we do not usually pay much attention to them. For example, when we read a book we must make decisions at several levels:

(a) recognition of individual letters,
(b) deciding whether the text is meaningful, (recognising words)
(c) understanding the text, (deciding which concepts are being discussed)
(d) deciding whether the message is relevant.

This book is concerned with making decisions in a computer or electronic equipment. The type of decisions that our machines can make are exemplified by (a); decisions such as those listed in (c) and (d) are more complex and outside the range of our discussion.

It is important to note that recognition demands decision. We *recognise* a mark on a page as a letter "A" by *deciding* that it is more like some mental image of "A" rather than "B" or "C". By analysing an event, and placing a label on it, we are effectively *classifying* it. The terms, recognition, detection, decision-making, classification are virtually synonomous although they do have slightly different shades of meaning:

recognition is quite specific. For example, we recognise a well known face in a crowd. Notice that the object being recognised is known beforehand.

detection is less specific than recognition and we may not have prior detailed knowledge of what it is that is being detected. We are able to detect the presence of a signal in noise even though we have no detailed picture of what a signal or noise is.

classification is used to describe those tasks where the groups being separated are diffuse and ill-defined. For example, we should classify animals into species.

decision-making has the wider connotations implied by the four examples cited above. However, in this book the term will be used synonomously with *classifying*.

In this book we shall discuss *classification* rather than the broader aspects of decision-making outlined above. A device for classifying events will be called a *classifier*. This may be a computer program, an electronic machine, or the mathematical rules defining how such a machine behaves.

1.2 MOTIVATION

Now that we have discussed, in abstract terms, what our machines can, or should do, it seems appropriate to list a few of the practical tasks that we can envisage as being within the scope of our subject. From this list it will be seen that there are several ways in which the construction of intelligent machines for decision-making could help men to a better life. The following is not intended to be a complete list of applications* but is merely intended to convey the great rewards that are possible if we can successfully solve the problems of pattern classification:

(a) automatic reading of cheques, invoices and postal codes,
(b) automatic medical diagnosis and prognosis. In some cases it might even be possible to control treatment automatically,
(c) classification of biological species for ecological studies,
(d) detecting or predicting failures in expensive machines,
(e) detecting atmospheric or water pollution.

It must also be admitted that there are many applications that would help men to wage war more effectively, but as in every other field of study, this should not deter us from seeking to develop the techniques for more worthy causes.

The major benefits that we can foresee for a successful solution to these problems are:

(a) relieving men and women from tedium; giving them the opportunity to devote more attention to interesting work,
(b) eliminating danger and/or unpleasant duties,
(c) achieving more consistent results than is possible by human means,
(d) saving money,
(e) allowing some of the skills of the most experienced professional people to be available cheaply and universally. Of course, we do not conceive of a situation in which the computer could totally replace a skilled doctor, although in certain restricted cases (for example in differential diagnosis) automation could possibly achieve results comparable with the best doctors,
(f) improving the quality of manufactured products.

* A more complete list is given in Chapter 7.

1.3 RELATIONSHIP TO OTHER FIELDS

Pattern classification is related to a number of other fields including,

control engineering,
taxonomy,
statistics,
numerical analysis,
logical design,
systems engineering.

The detailed comparison between these subjects and our own, is best delayed until we have described the basics of pattern classification (section 2.9).

Pattern recognition workers have been quite willing to learn from any field that may be relevant to their interests. As a result, there has been a long discussion between physiologists and psychologists on the one hand and engineers on the other. It was felt by all of these workers, that the study of men and animals might help in the development of machines for pattern classification. This approach led to some of the earlier ideas which have now been surpassed by more powerful techniques for classification. The major benefit of this discussion has been the development of measurement systems for machines that can read. It is rather easier (that is less difficult!) to work at the lower levels of the central nervous system, and some of the most significant advances have been made in the discovery of primitive visual sensors capable of detecting edges, lines and corners. Our subject is more closely related to the kind of tasks that we imagine are performed in the cerebral cortex. Since it is so difficult to find out anything about this region of the brain, very little benefit has been felt in the engineering world from this discussion.

1.4 OUTLINE OF THE BOOK

Chapter 2 lays the basic foundation for the rest of the book. Chapter 3 describes over 20 distinct methods of classifying patterns. The newcomer to the subject can safely omit Chapter 4 which compares these classifiers. From this large number of suggested classification schemes, we choose three that are worthy of further consideration. Chapter 5 describes methods for calculating the parameters necessary to achieve the required classification. Chapter 6 lists the practical details that will help the reader to obtain the best and most reliable results. This chapter also describes how we can build a pattern classifier in analogue or digital equipment. Chapter 7 describes some of the applications of pattern classification, and lists some of the results that have been obtained using the techniques described earlier. In the final chapter we discuss a number of problems which do not fit into the foundation laid in Chapter 2.

The author feels that in the future we should develop several fundamentally different models, since there are several distinct types of pattern classification task.

Appendix 1 presents a flow chart which defines the practical steps required in the design of a pattern classifier. Appendix 2 discusses a number of data analysis techniques which can help us to learn more about the data before we start to design a classifier. These provide several benefits, including the ability to test our results against the best possible classifier. Appendix 3 describes techniques for the generation of synthetic data and its use in testing pattern classification procedures. Appendix 4 discusses one of the most recent developments in the subject.

Chapter 2

The Basic Model

2.1 INTRODUCTION

In this chapter we shall lay the conceptual foundations for the rest of the book. To do this we must use the shorthand notation of mathematics. Even though our approach to the subject does not require an advanced mathematical knowledge, a working familiarity is certainly valuable.* We must also develop the vocabulary of high-dimensional spaces, but this can be achieved by drawing parallels to the familiar two and three-dimensional equivalents.

Our approach is heuristic, pragmatic, and experimental. This is in sharp contrast to most other authors [1, 2] who adopt formal mathematical methods of analysis. We have two rules:

(a) If a technique is useful *and* easily implemented (either in a computer or special-purpose electronics) then we shall use it.
(b) If a procedure does not satisfy the second of these conditions then we shall shelve it until it can be implemented rather more easily than at present.

We must make it clear that any formally-based techniques, which are useful in the practical problems that face us, will be welcomed *and used.*

2.2 PATTERN DESCRIPTION OR ANALYSIS

Pattern recognition requires at least two processes:analysis (alternatively called feature extraction, pattern description or measurement) and classification. The boundary is ill-defined, although we must be careful to specify exactly what kind of interface is required between the analysis and classification stages.

Consider Figure 2.1. The analyser accepts an input called the *pattern*, which is a very complex physical event, such as a television picture of a typed page, or the signal from a microphone. Patterns may be optical, acoustic, electrical, numbers on a page, a radar signal, a biochemical profile of a patient. The list is seemingly endless. It is unlikely that any part of the analyser can be a general-purpose device.

* We use nothing more advanced than the summing operator and the logical connectives.

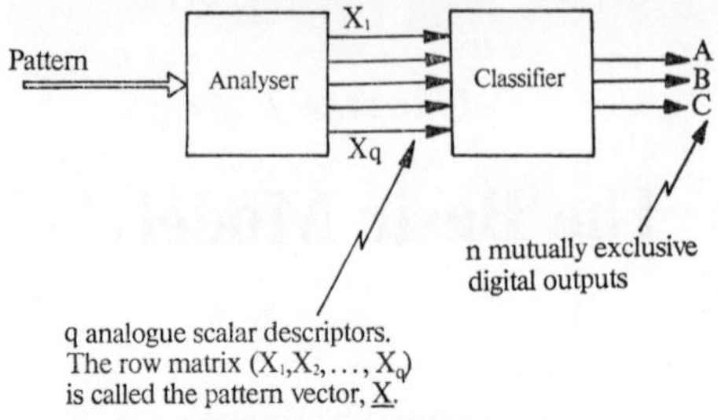

Figure 2.1 A pattern recognition system incorporates two distinct processes: analysis and classification.

2.2.1 The Analyser Output, the Pattern Vector

It is usually quite convenient to describe a pattern by taking several parallel measurements. The numeric values obtained in this way will be called *descriptors*. It will be convenient to allow a number of descriptors to be used simultaneously to describe the pattern. Since a descriptor is only a number, it is essential to be able to associate it with a particular observation. We shall therefore find it convenient to describe the pattern by a vector whose terms are the descriptors. This is called the *pattern vector* or *descriptor vector*. It is a perfectly ordinary vector with no complications, but to be absolutely certain that we establish the correct picture we shall list the essential features:

(a) It has a fixed number of elements, called descriptors.
(b) Values for all descriptors are always known, and are always numeric.* True and false; married and single; small, medium and large are also allowed, since they can be coded in numeric terms. For example: small = 1, medium = 2, large = 3.
(c) The order of descriptors within the pattern vector is fixed.

The following are admissible pattern vectors:

$$(196.8, -2.74, -1, +3),$$
$$(-3, 0, 1),$$
$$(0, 1, 1).$$

The following are not admissible pattern vectors:

(2.6, −3, green),
(3.7, −2.9, ?) (? is "don't know").

The following pattern vectors are different, since the order is changed:

$$(-1, +2, -3),$$
$$(+2, -1, -3).$$

* Numeric is taken to include cardinal, ordinal, fractional or logical quantities.

2.2.2 Descriptor Range Normalisation

It will be convenient to have a pattern vector whose components are all in the same range. The range is quite arbitrary, and we need not feel that it is essential to have exactly the same range for all of the descriptors. Normalisation will allow us to develop simplified hardware and computer programs.

2.2.3 Information Destruction in the Analyser

The input to the analyser will normally have a very high information content, possibly in excess of 10^5 bits/pattern.* Our pattern recognition system may be designed to select one of 2 to 100 outputs, representing an output information content of 1 to 7 bits. The major function of the analyser is information destruction. This process should be selective, keeping relevant information, but discarding any that will not assist in the classification. Almost always, some of the useful information will be lost in the analyser. This will cause problems, since it may mean that two classes of patterns which were originally quite distinct have now become partially indistinguishable. This will have repercussions throughout this book, and will be discussed more fully in a subsequent section of this chapter.

2.2.4 The Problem of Designing the Analyser

It is important to realise that it is extremely difficult to suggest and select good descriptors. So far there is no really satisfactory solution to this, which is thought by many workers to be the central problem of pattern recognition. The present analyser designs are only partially successful. Even this limited success has been obtained by using human intuition, since there are no really effective methods for selecting descriptors automatically [3, 4].

The problem of automating the search for good sets of descriptors is difficult, because it is impossible to define a useful figure of merit for each descriptor considered alone. It is possible to do this for a complete set of descriptors. An exhaustive search of all possibilities is quite out of the question. Suppose we have 100 suggested descriptors among which there are only 10 useful ones. Then we should have to investigate at least

$$\sum_{r=1}^{10} {}^{100}C_r$$

sets of descriptors to find them by an ordered search.†

* This value was calculated for an optical system with 400 × 400 picture elements, each having 10 brightness levels. In many problems we find that the information is in an analogue form. It is then impossible to be precise about information levels.

† This number is approximately $2.5.10^{14}$.

Fortunately the situation is often saved by human intellect. It is often possible for a man, who has worked with a particular kind of problem for many years, to define useful measurements. Often it is guesswork that leads to success! Before we attempt to design a classifier, we shall assume that (somehow) the analyser has been designed. We hope to convince the reader that this assumption is not totally unrealistic, by presenting some examples later in this chapter and in Chapter 7. For the moment, we must ask the reader to accept this on trust, until we have completed our description of the basic model.

2.2.5 Notation Used for Pattern Vectors

We shall assume that there are q descriptors which will be denoted by

$$X_1, X_2, \ldots, X_q$$

The vector

$$(X_1, X_2, \ldots, X_q)$$

is the pattern vector and will be denoted by $\mathbf{X}$. In the text, bold-face type denotes a vector, while underlining is used in the diagrams.

2.3 THE CLASSIFIER

The input to the classifier is the pattern vector, $\mathbf{X}$. Its output is a series of digital signals which are mutually exclusive and exhaustive. The function of the classifier will become more obvious when we discuss the overall system in a later section. For the moment, we shall simply say that it is a device for selectively destroying information. A typical classifier might have an input information flow-rate between 20 and 1000 bits/pattern, and an output information flow-rate of 1 to 7 bits/ pattern.

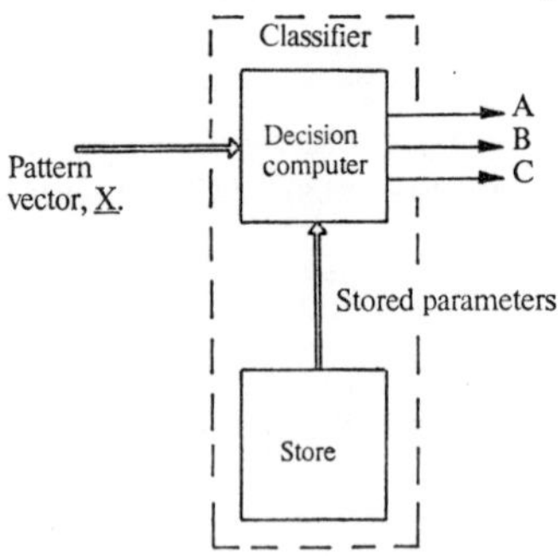

Figure 2.2 The classifier contains a store and a function generator, called the decision computer.

The classifier may be further subdivided into two sections (Figure 2.2):

(a) a store,
(b) a device for making decisions.

We must find a suitable structure for the decision computer as well as obtaining values for the stored parameters. The former will be discussed at length in Chapters 3 and 4, while Chapter 5 concentrates on the estimation of parameters.

2.4 THE TEACHER

For the uninitiated reader, the title of this section is, perhaps, a little surprising. The concept of a teacher is introduced to help us to define and discuss the problem of pattern recognition. During the design phase for our machine, the teacher examines the same pattern as the analyser, and produces a response (equivalent to the output of the classifier). If our pattern recognition system were perfect, it would produce the same response as the teacher, for all input patterns.

Why do we try to build a machine when, apparently, we already have a teacher, which can perform the desired task? In some cases the teacher has no physical existence at all; in other situations the teacher is too expensive to use all the time, or is too slow. However, the best way of answering this question is to consider one particular example, medical diagnosis. At a post-mortem examination it may be possible to find the cause of death, but it would be an enormous advantage to be able to find the pathology before the patient dies. It may even save the patient's life! Now, we might be able to design a pattern recognition system which could *predict* the post-mortem findings by using measurements that are taken on a *living patient*. If we were successful, we could anticipate what the results of a post-mortem examination would be, if the patient were to die, and possibly avoid that happening. In this optimistic example, the teacher is the pathologist and the teacher output is his post-mortem report.

The teacher is not always a recognition device, for example, the teacher output may be the answer given by a person when he is asked what alphabetic letter he has just written.

The important point is that the teacher is a useful concept. It is the ideal recognition device and defines the correct classification of the input pattern.

2.4.1 Notation

Throughout this book we shall use T to denote the output of the teacher.* The output of the classifier will be denoted by M.

* The following terms are also used to describe T: required, desired, or correct classification.

2.4.2 Restriction to 2 Classes

It does not seriously restrict the usefulness of our techniques if we limit our discussion to problems with two classes. Multi-class problems will be considered in sections 5.13 and 6.2.10, but in the remainder of the book our teacher and classifier outputs can assume only two values, ± 1. Multi-class recognition devices can of course be produced by combining the outputs of a number of 2-class systems which are placed in parallel (Figure 2.3).

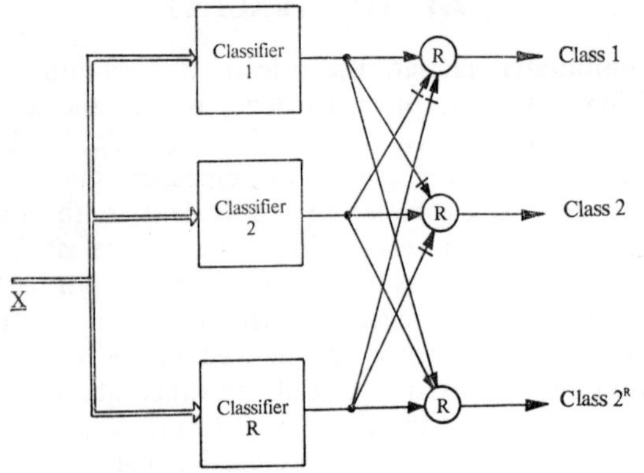

Figure 2.3 Construction of a multi-category classifier by logical combination of the outputs obtained from a bank of two-class devices.

2.4.3 Learning by Example

The problem of pattern recognition arises because we do not know how to discriminate between ill-defined sets such as the classes of hand-written "A" s and "B" s. If we ask a dozen people how they can discriminate between these two groups, we shall probably obtain twelve different answers. Our subjects will almost certainly describe criteria which are based on such indefinables as "straight lines", "bends" and "corners" (these are not the geometric figures, but are more like crude sketches). The same kind of difficulty is encountered in other pattern recognition problems; indeed this is the very essence of our subject. The reader should have no illusions that pattern recognition is a precise subject. It is part of a growing field, which might be called the art of cheap and cheerful computation.

Since we cannot define an "A" or "B", how can we ever hope to recognise them by machine? We can, of course, collect together a number of examples of each class, and use these to design the recognition equipment. This process is a direct parallel of the way children learn. We cannot tell a child *how to speak*, but he will *learn from our example*. We shall require that the patterns are all labelled by the teacher, but that no other information be available. In particular, we shall not demand that the teacher be able to indicate how confident he is about his decision, since we cannot always rely on this information being available.

We are left then with the simple requirement, that the teacher labels every pattern with either of the two permissible outputs: $T = -1$, or $T = +1$. We shall study techniques for designing the classifier so that its output, M, is an accurate predictor of T. Our design techniques are automatic; once they have been started they do not require human intervention.

2.4.4 The Deity of the Teacher

Consider Figure 2.4. How many "A" s are drawn? The author feels that there are two, but some people will contend that there are more,

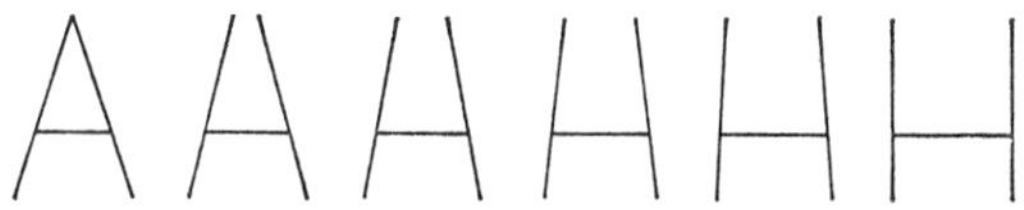

Figure 2.4 How many A s are drawn here? Some people will regard certain characters as A while others will feel they are more like an H. Hence it is impossible to obtain a teacher whom everybody regards as always being correct.

or fewer. There is no *correct* answer; there cannot be any absolute truth in matters where subjectivity is concerned. If the author were to act as a teacher for a pattern classifier, there would probably be objections from other people that he is not a very good one. These other contenders for the rôle of teacher would have a perfect right to criticise the author, but no right at all to criticise the classifier, since this device is designed with the author acting as the supreme authority. (Don't criticise the child, he only did as his parents taught him!)

As far as we are concerned here, the teacher is the supreme authority. The teacher defines the classification of the patterns, and it is the decision of the teacher that we are trying to emulate. The choice of a suitable teacher, whether this is an individual or consensus opinion, is not part of our concern, but belongs to the worker who is more familiar with the application area. For example, the choice of a teacher for a problem of medical diagnosis is a matter for discussion among doctors.

2.5 VECTORS AND SPACES

We could not have designed the learning techniques described in Chapter 5 if we had not been able to represent information in visual form. Although we shall present mathematical descriptions of our methods, they do not give any sort of "feel" for the problem. To do this, we need some form of presentation which will allow us to think about

Table 2.1 Data for a very simple problem in pattern classification. We are given two descriptors (X_1 and X_2) on each of 30 patterns. (10 samples of class 1, 10 of class 2 and 10 of unknown classification.) We are required to indicate the more probable classification for each of the unknowns. Answers are given in the legend to Figure 2.5.

Class 1		Class 2		Unknown	
X_1	X_2	X_1	X_2	X_1	X_2
145	64	168	66	150	68
112	65	117	69	106	63
118	64	133	70	130	62
126	62	145	72	133	63
128	64	134	69	147	65
126	66	154	68	196	67
152	63	202	71	117	65
136	66	162	68	133	66
120	65	150	70	127	69
118	62	167	63	146	71

pattern classification in visual terms. To emphasise this point, we shall consider a simple pattern classification problem which, incidently, has most of the ingredients of a more nearly life-sized application.

Consider Table 2.1, but for the moment do not plot any graphs. We are given ten samples from class 1, ten from class 2 and ten which

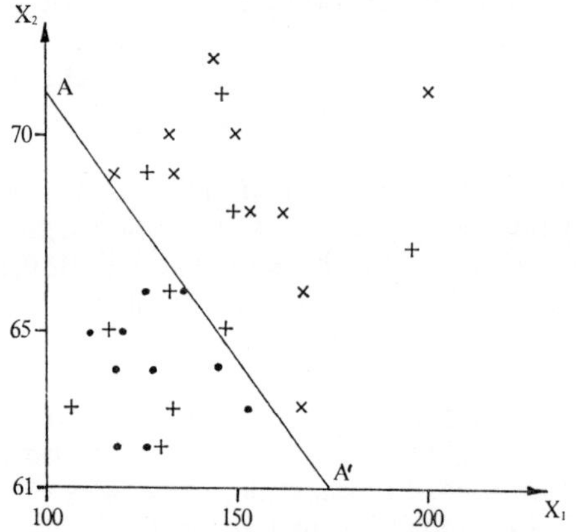

Figure 2.5 Presenting the same data as Table 2.1 in graphical form to emphasise that graphical information can be more easily assimilated than tabular information. The data was obtained by measuring weights (X_1) and heights (X_2) of women (class 1, circles) and men (class 2, diagonal crosses). The classification of the unknown (upright crosses) may be accomplished by finding which side of the line AA' they fall. Those points above the line are assumed to be men; those below, women. The true classification of the unknowns listed in Table 2.1 are:

2, 1, 1, 2, 2, 1, 1, 1, 2, 2.

we are asked to attribute to the more appropriate class. It is difficult to find a criterion for separating the two classes, even though there are only two descriptors. It is much easier to assimilate this information in graphical form (Figure 2.5). Here we have plotted X_1 and X_2 along the co-ordinate axes of a two-dimensional space. It is easy to allocate the unknown points to their appropriate classes by simply finding whether they lie above or below the straight line. (For the moment we must defer any discussion on how this line was obtained.) Suppose our patterns each had 3 descriptors, we should then require a space with 3 dimensions to represent the point unambiguously (Figure 2.6). It is

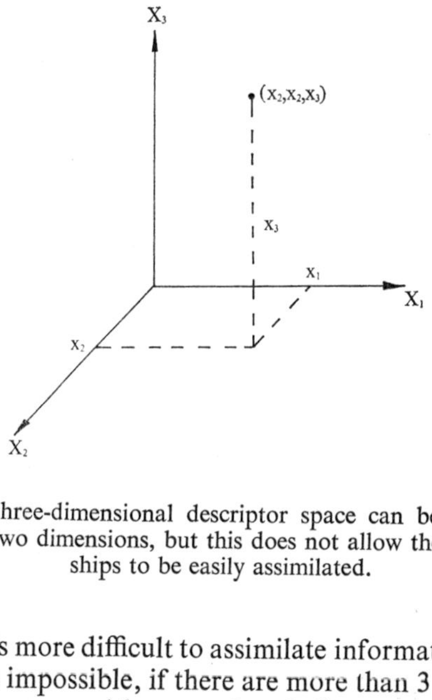

Figure 2.6 A three-dimensional descriptor space can be presented by a projection into two dimensions, but this does not allow the spatial relationships to be easily assimilated.

apparent that it is more difficult to assimilate information in a 3-dimensional space. It is impossible, if there are more than 3 descriptors, since we cannot draw spaces with more than 3 dimensions.

Nevertheless the terminology applicable to 2 and 3-dimensional spaces can be readily extended to cover those situations in which there are more than three descriptors. For example, a straight line in two dimensions (X_1, X_2) has an equation of the form:

$$W_1X_1 + W_2X_2 = F \tag{2.1}$$

where W_1, W_2 and F are constants. The equation of a plane in a 3-dimensional space has the form;

$$W_1X_1 + W_2X_2 + W_3X_3 = F \tag{2.2}$$

Table 2.2 Summary of the terms used in the geometry of high-dimensional spaces

2 dimensions	3 dimensions	q dimensions	Comments
2-dimensional space, 2-space	3-dimensional space, 3-space	q-dimensional space, *q*-space, Hyperspace, **X**-space, Space	Terms are used synonomously. Co-ordinate axes are the components of the vector: $\mathbf{X} = (X_1, X_2, \ldots, X_q)$
Curve	Surface	Hypersurface, Surface	Terms are used synonomously.
Straight line	Plane	Hyperplane	Equation is typically $\sum_{i=1}^{q} W_i X_i = F$
Circle	Sphere	Hypersphere	Equation is typically $\sum_{i=1}^{q} (W_i - X_i)^2 = F$
Square	Cube	Hypercube	—
Ellipse	Ellipsoid	Hyperellipsoid	—
Polygon	Polyhedron	Polyhedron	Convex hull of at least $(q + 1)$ points distributed at random in a q-space
Vector, point			Used synonomously although a point is, strictly speaking, the tip of the vector. X is a typical vector where: $\mathbf{X} = (X_1, X_2, \ldots, X_q)$
Distance (Euclidean)			Distance between vectors $(U_1, \ldots, U_q)$ and $(V_1, \ldots, V_q)$ is $\sum_{i=1}^{q} (U_i - V_i)^2$

When there are more than three dimensions we speak of a *hyperspace* and the equivalent mathematical expression, viz:

$$\sum_{i=1}^{q} W_i X_i = F \qquad (q \text{ is the number of dimensions}) \tag{2.3}$$

defines a hyperplane in it. If a point $(X_1, X_2, \ldots, X_q)$ satisfies (2.3) it is said to be on the hyperplane. The terminology is exactly the same as we use for a 3-dimensional space, with the addition of the prefix *hyper-*. Table 2.2 summarises the terminology of high-dimensional spaces [5, 6].

2.5.1 Decision Surfaces

We have already met a simple decision surface in Figure 2.5. In this example it was possible to separate the sets using a straight line.* If

* If we had used larger samples, we should have found that the classes "men" and "women" cannot be separated using only weight and height.

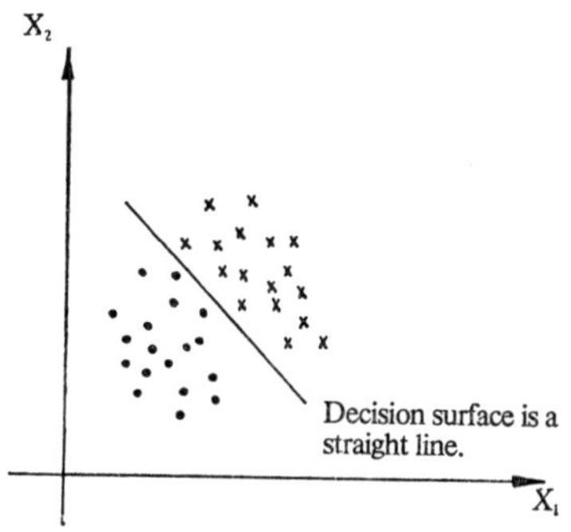

Figure 2.7 Linearly separable sets in a two-dimensional space.

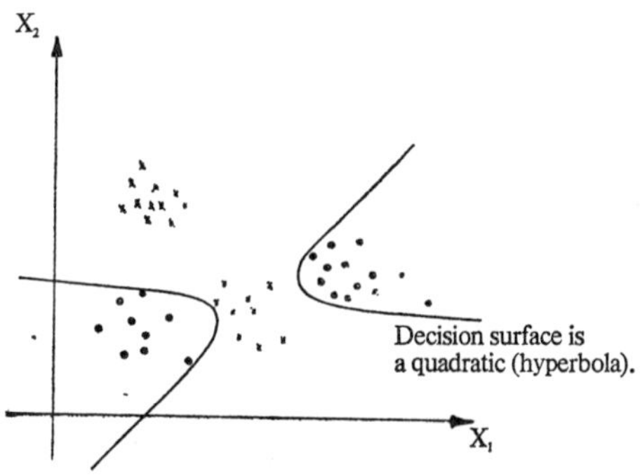

Figure 2.8 A quadratic surface would be required to separate these sets.

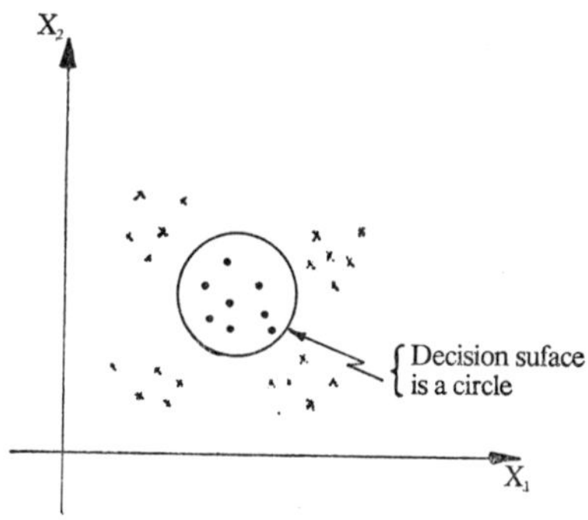

Figure 2.9 Another situation requiring a quadratic decision surface to separate the classes of circles and crosses. This type of distribution of the points occurs in certain medical problems. Circles might correspond to "normality" and crosses to "disease".

they can be separated by a straight line the sets are said to be *linearly separable*. (In a high-dimensional space the sets would be linearly separable if there existed a hyperplane which could discriminate between them.) Figure 2.7 shows another example in which the decision surface is linear. Not all pairs of sets can be separated using such simple surfaces. For example, quadratic surfaces would be required in some

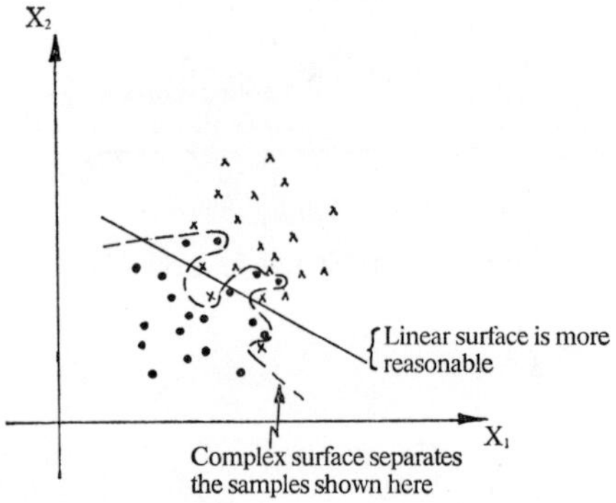

Figure 2.10 Two finite samples from overlapping classes are shown. While it is possible to separate these samples using a complex surface, it is apparent that a simpler (linear) surface would be more reasonable. This linear surface will perform better on the parent populations.

situations (Figures 2.8 and 2.9). In some situations it is not possible to divide the (infinite) classes by any finite curve, although any finite samples from these classes can always be separated by a curve of sufficient complexity (Figure 2.10).

A pattern classifier is a device for partitioning the space defined by the analyser outputs (**X**-space). That is, it constructs a decision surface, which we hope can be placed in such a position that it separates the classes defined by the teacher. Suppose, for the moment, that this is possible. When its decision surface is placed in this optimal position, the classifier and teacher always agree. In this case we may regard, the teacher as making decisions by the use of a decision surface in **X**-space.

In the situations when the classes are not separable in **X**-space, we cannot draw a decision surface for the teacher. We can, however, still draw a decision surface for the classifier.

2.5.2 Similarity, Distance and Compactness

Two patterns will produce identical **X**-vectors, if the patterns are identical. (The reverse is not always true.) If two patterns are similar then we shall *usually* find that their corresponding **X**-vectors are close together.This is not a universal rule, but a broad generalisation, which indicates that the analyser contains a finite number of discontinuities.

The distance between two points in a high-dimensional space can be defined in an infinite variety of ways. The usual definition of distance in two and three-dimensional spaces can be extended to higher dimensional spaces. If $(U_1, U_2, \ldots, U_q)$ and $(V_1, V_2, \ldots, V_q)$ are two points in the space then the Euclidean Distance between them is defined as:

$$\sqrt{\sum_{i=1}^{q} (U_i - V_i)^2} \tag{2.4}$$

Let **U** and **V** be two q-dimensional vectors. Any quantity, $D(\mathbf{U}, \mathbf{V})$, qualifies being called a distance function if it satisfies three conditions:

$$\left.\begin{aligned} &D(\mathbf{U}, \mathbf{V}) = 0, && \text{if and only if } \mathbf{U} = \mathbf{V} \\ &D(\mathbf{U}, \mathbf{V}) > 0, && \text{if } \mathbf{U} \neq \mathbf{V} \\ &D(\mathbf{U}, \mathbf{W}) + D(\mathbf{W}, \mathbf{V}) \geq D(\mathbf{U}, \mathbf{V}), && \text{if } \mathbf{U} \neq \mathbf{W} \neq \mathbf{V} \end{aligned}\right\} \tag{2.5}$$

On this basis all of the following are distances:

$$\left.\begin{aligned} &\left\{\sum_{i=1}^{q} (U_i - V^r)\right\}^{1/r} && \text{Minkowski } r\text{-distance} \\ &\sum_{i=1}^{q} |U_i - V_i| && \text{City block distance} \\ &\max_{i=1,\ldots,q} \{|U_i - V_i|\} && \text{Square distance} \end{aligned}\right\} \tag{2.6}$$

These will recur again, in Chapter 3, but we hope that we have convinced the reader that the definition of distance is quite arbitrary. Similarity is also an ill-defined quantity; it is difficult to think of any measure for similarity other than subjective assessment. For the moment, we shall assume that any two patterns within the same class are "similar", and that they are "dissimilar" if they belong to different classes. Then our (vague) rule states that we are likely to find that the distances between the vectors corresponding to patterns from the same class are small. That is, they are likely to be small compared to the distances between patterns from different classes.

This is a naïve view that we must develop further. Let us consider type-written letters as our patterns. If we look very closely at such characters beneath a magnifying glass we see that there are some remarkably large differences between fonts (Figure 2.11). Now it would be unreasonable to expect all "A" s to be mapped into the same small region of the **X**-space by the analyser. We shall probably obtain a cluster of points corresponding to each font (Figure 2.12). It is not possible to define a cluster in any convincing way;* the reader's own

* A cluster corresponds to a peak in the probability density function.

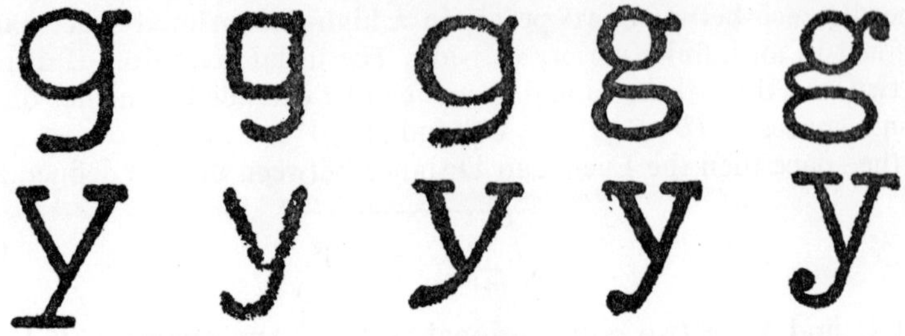

Figure 2.11 Showing the range of type fonts in common use.

mental picture of a cluster, swarm or galaxy is much more satisfactory. A class may contain several well-separated clusters. (Figure 5.33B shows an artificial class with 3 clusters.) In such classes the inter-cluster distances are large compared to the intra-cluster distances. Sometimes co-classed clusters will overlap. (Figure 5.33A also contains 3 clusters.)

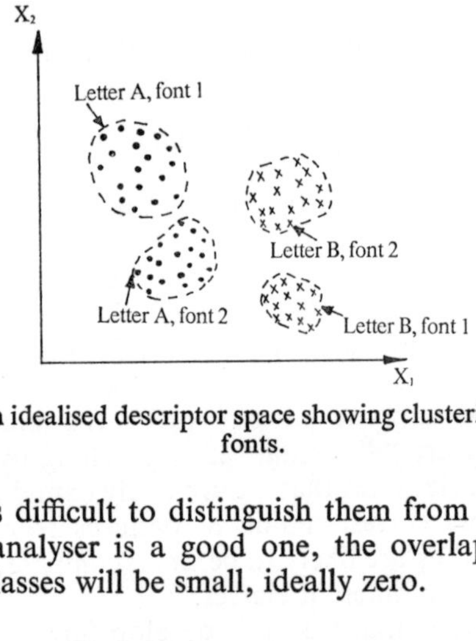

Figure 2.12 An idealised descriptor space showing clustering due to different fonts.

In this case it is difficult to distinguish them from a single elongated cluster. If our analyser is a good one, the overlap between clusters from different classes will be small, ideally zero.

2.6 TRAINING AND TEST SETS

We shall assume, throughout this book, that the analyser has already been designed. Hence we expect our input data to consist of a number of pattern vectors. It is wise to split this data into two parts. One can be used to design the classifier, while the other is used to evaluate the performance of the system after the design is complete. That set of data used for designing the classifier will be called the *training set*, and the other, the *test set*.

The use of distinct training and test sets eliminates the possibility of quoting over-optimistic results. This could occur, if we used the same small set of data for the design and evaluation of the classifier. (It is far less likely to be a serious problem if the data set is very large.) The reason for our obtaining over-optimistic results may be understood with the help of Figure 2.10. Suppose we were given the points shown here to design and evaluate our classifier. If we were not very careful, we might produce a decision surface, like the broken curve, which will discriminate perfectly between the given samples of the pattern classes. However, it will not produce such a good performance when it is set to work on new samples.

The given data should be sorted into the training and test sets using a random selector. This will remove any bias due to inter-pattern correlation.

2.7 EXAMPLES OF REAL DATA

So far we have discussed the data format in an abstract way. It will be worthwhile spending a brief time here to show that our ideas bear some resemblance to the real data that we hope to analyse. We shall briefly discuss the following examples:

(a) taxonomic classification of Iris,
(b) discrimination between handwritten "A" s and "H" s,
(c) diagnosis of disease. Three examples are given,
(d) discrimination between the spoken consonants /s/ and /f/.

At this stage we merely wish to convince the reader that the model we have described is a reasonable one. Since we are going to discuss real problems in very few words we must inevitably simplify them to their essential features.

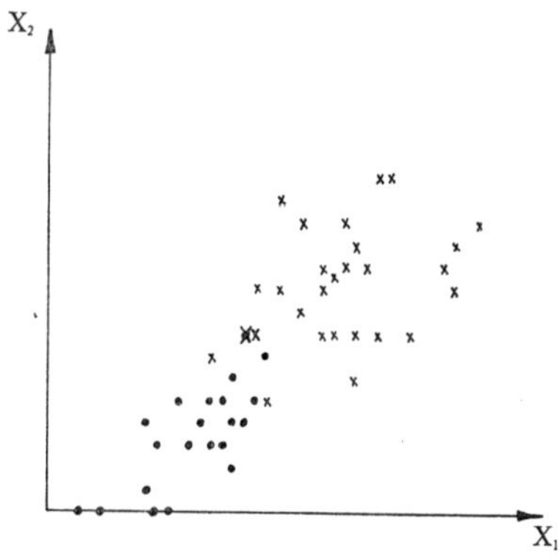

Figure 2.13 A simple taxonomic problem expressed in terms of a descriptor space. The co-ordinate axes are: X_1, petal length, X_2, petal width. The categories are: circles, Iris Versicolor, crosses, I. Virginica. A third species, I. Setosa, which is not shown, produces a compact cluster, situated deep in the third quadrant of this space.

2.7.1 Classification of Iris

R. A. Fisher's data on irises has been the subject of many different attempts to analyse it [7]. The data consists of the following measurements taken on 50 samples of *I. setosa*, *I. versicolor*, *I. virginica*:

petal length,
petal width,
sepal length,
sepal width,

Of these we have selected the first two and have plotted a scattergram (Figure 2.13). It is clear that *I. virginica* and *I. versicolor* can be separated reasonably well on the basis of these two measurements alone. *I. setosa* is even more easily recognised, since it produces a compact cluster well away from the other two.

2.7.2 Discrimination Between Handwritten "A" s and "H" s

Photographs of some of the samples used in this study are shown in Figure 2.14. The letters were written in squares (1.5″ × 1.5″) with a felt-tipped pen. The square was quantised on to a 40 × 40 array of elements each having 64 brightness levels. From this array the central

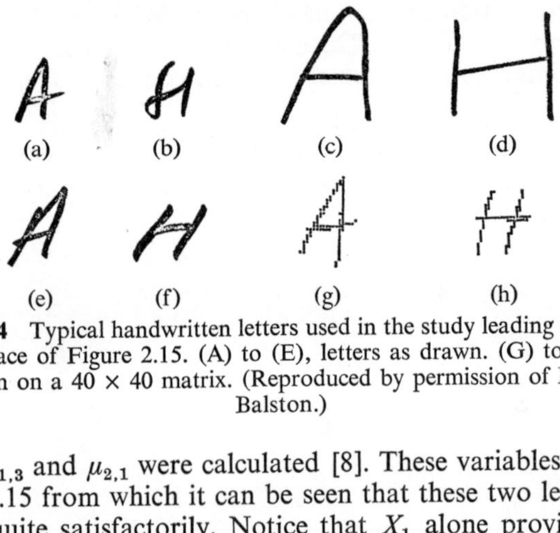

Figure 2.14 Typical handwritten letters used in the study leading to the descriptor space of Figure 2.15. (A) to (E), letters as drawn. (G) to (H) after digitisation on a 40 × 40 matrix. (Reproduced by permission of Dr. D. M. Balston.)

moments $\mu_{1,3}$ and $\mu_{2,1}$ were calculated [8]. These variables are plotted in Figure 2.15 from which it can be seen that these two letters can be separated quite satisfactorily. Notice that X_1 alone provides a good basis for discriminating between these letters.

2.7.3 Diseases of Calcium Metabolism

Figure 2.16 is a restatement of data presented by Fourman [9, Table 4.3]. (Since Fourman gives no quantitative data, we have assumed that *low* may be represented by zero units in our descriptor space; *normal* by 1 unit; *high* by two units. This coding may distort the space slightly.) We have not investigated this problem in any detail, so we do not know

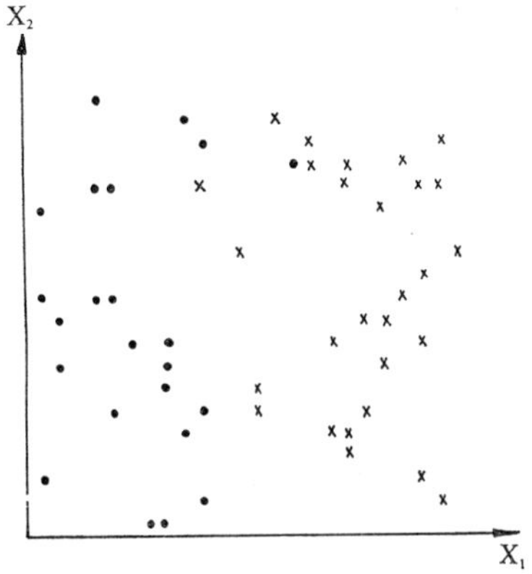

Figure 2.15 A descriptor space obtained from handwritten characters. X_1 and X_2 are central moments of the characters, which were drawn with a felt-tipped pen within a 1.5 inch square. X_1 is the moment $\mu_{2,1}$, X_2 is the moment $\mu_{1,3}$. Crosses are H s and circles are A s. (Reproduced by permission of Dr. D. M. Balston.)

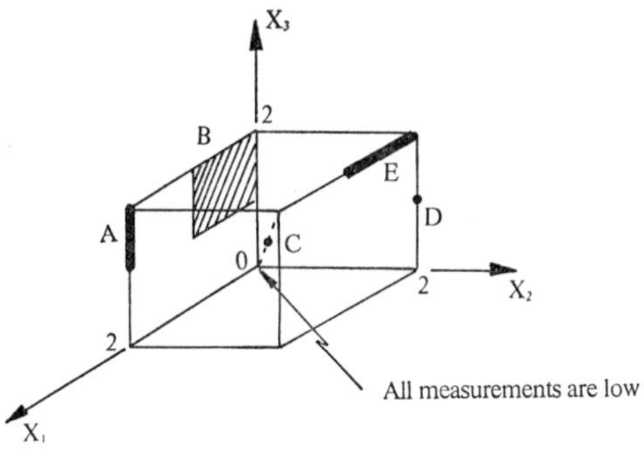

Figure 2.16 Showing the spatial relationship between diseases of calcium metabolism. The co-ordinate axes are: X_1, serum calcium; X_2, serum phosphate; X_3, alkaline phosphatase. Categories are: A, parathyroid excess; B, rickets and osteomalacia; C, osteoporosis (bone atrophy) and the normal group, which overlap; D, parathyroid deficiency; E, renal osteodystrophy. The diagram shows typical members for each category, which would be more accurately represented by diffuse clouds of points.

to what extent the classes overlap in this space. Our reason for presenting this example is that it relates a diagnostic task to a descriptor space; no other inference should be made.

Notice that, like Figure 2.9, there is a region corresponding to normality which is surrounded by disease groups.

2.7.4 Disorders of the Acid-base Balance

Figure 2.17 shows the spatial relationship between four disorders of the acid-base balance in the human body. This diagram presents the same data as Baron [10, Figure 2.2] and Houston *et al.* [11, p. 384]. Notice the central region of "normality" surrounded by the four disease categories.

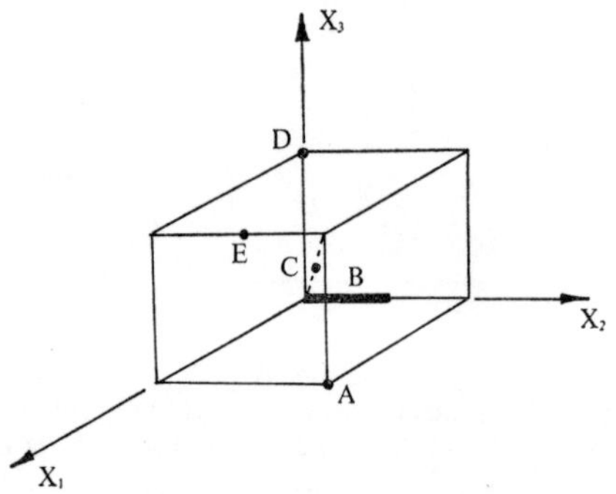

Figure 2.17 Showing the spatial relationship between four disorders of acid-base balance. The co-ordinate axes are: X_1, plasma bicarbonate; X_2, partial pressure of carbon dioxide in plasma; X_3, plasma pH. The categories are: A, respiratory acidosis; B, metabolic acidosis; C, normal; D, respiratory alkalosis; E, metabolic alkalosis.

2.7.5 Kidney Disease

Another medical example is presented in Figure 2.18. This shows the same kind of distributions that we saw in the two previous cases. This data was derived from Houston *et al.* [11].

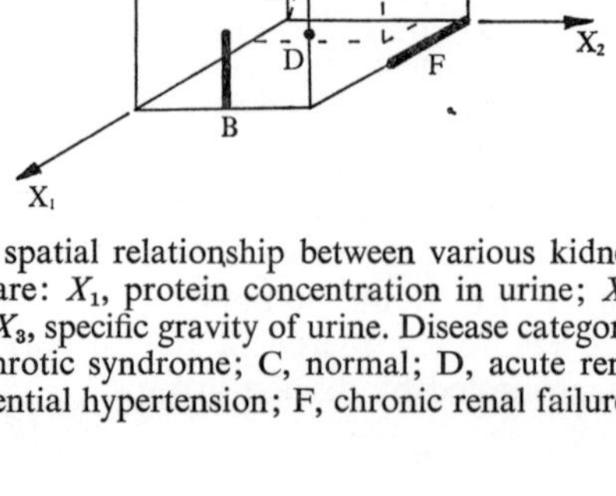

Figure 2.18 The spatial relationship between various kidney diseases. The co-ordinate axes are: X_1, protein concentration in urine; X_2, daily urinary output (volume); X_3, specific gravity of urine. Disease categories are: A, acute nephritis; B, nephrotic syndrome; C, normal; D, acute renal infection; E, essential hypertension; F, chronic renal failure.

2.7.6 Speech Recognition

Figure 2.19 is a reproduction of a diagram published by S. H. Lavington [12]. This simple space has co-ordinate axes given by the

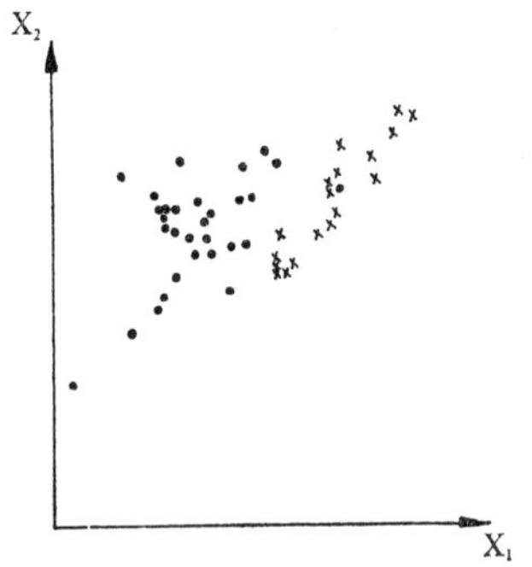

Figure 2.19 A measurement space derived from speech signals. X_1 represents the number of zero crossings/unit time. X_2 is the number of turn-arounds (stationary points) in the same unit of time. Crosses represent samples of the initial consonant /s/, and circles the initial consonant /f/. (Reproduced by permission of the Institution of Electrical Engineers.)

number of zero-crossings (X_1) and turn-arounds (X_2) in the speech waveform in unit time. It is apparent that the two classes which represent the initial consonants /s/ and /f/ could be separated reasonably well in this space.

2.7.7 Data from Other Problems

The six examples that we have presented in this chapter are not typical of the great bulk of pattern recognition problems in two ways:

(a) There are usually many more than 3 dimensions.
(b) There are normally many more patterns provided in the training set.

A "typical" problem might require that we analyse 1000 patterns from each of 6 classes, where each pattern is coded by 20 descriptors.

2.8 SYNTHETIC DATA

Throughout this book we present experimental results to demonstrate the effectiveness of the learning schemes. Some of these experiments were conducted using data from problems of real practical interest. Others involved the analysis of data obtained from the physical world, although the discrimination problems we set ourselves were of little practical value. A third group of experiments was based on data which was produced using pseudo-random number generators [13, 14], and in some cases, a fixed classifier to simulate the teacher [15, 16]. The use of artificial data has a number of important features:

(a) It is cheaper than real data.
(b) By generating data during the execution of a learning program we avoid the use of large arrays for data storage. Most of our research can then be conducted using small computers.
(c) We retain complete control over the experiments. We can *define* an ideal solution. We can study problems of varying complexity; we can vary the number and position of clusters in the hyperspace, and the amount of overlap between classes.
(d) We can study problems in spaces with any number of dimensions. In particular we can use 2-space to gain the enormous advantage of visual interpretation of the results. We can draw scattergrams, and then compare these with the decision surfaces obtained by learning.

Three types of synthetic data have been used.

2.8.1 Type A Data

The simplest type of artificial data is produced using pseudo-random number generators and a fixed classifier to simulate the teacher (Figure 2.20). In our experiments we used 2 pseudo-random number generators

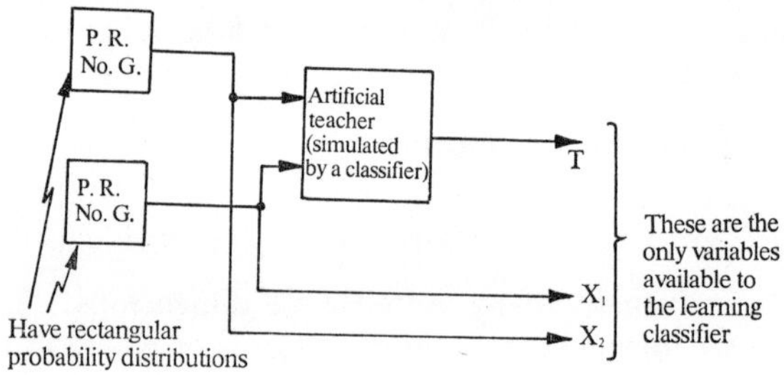

Figure 2.20 Method of generating type A synthetic data.

which had rectangular probability distributions. Hence the vector (X_1, X_2) uniformly scanned a square window in 2-space (Figure 2.21).

2.8.2 Type B Data

These experiments are more typical of the real problem than are type A. The pattern classes used in type B experiments are not separable and the probability distributions are not uniform. A 2-dimensional vector, **X**, and teacher decision, T, were generated as follows:

Let $\rho_{-1}(\mathbf{X})$ and $\rho_{+1}(\mathbf{X})$ be probability density functions, consisting of the union of a number of Gaussian clusters (Figure 2.22), [17].

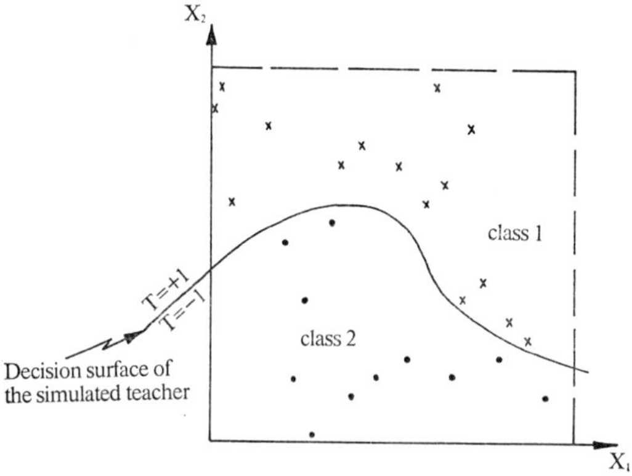

Figure 2.21 Type A synthetic data. The broken lines indicate the range of variation of the points (range of allowed patterns).

A pseudo-random binary sequence was used to generate T. An **X**-vector was then selected at random from a population with the distribution

$$\tfrac{1}{2}(1 + T)\rho_{+1}(\mathbf{X}) + \tfrac{1}{2}(1 - T)\rho_{-1}(\mathbf{X}) \tag{2.7}$$

The number and positions of the clusters is under our control. In addition the local spread around the modes can be varied at will.

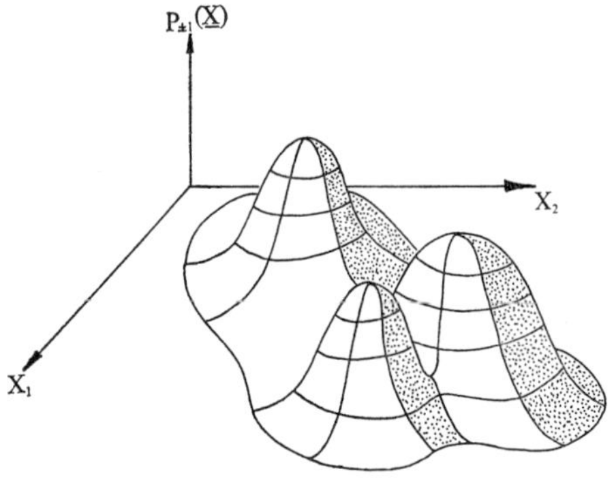

Figure 2.22 Type B synthetic data. A class (corresponding to $T = \pm 1$) is generated with a probability density function, $\rho_{\pm 1}(\mathbf{X})$, consisting of the union of a number of Gaussian hills.

2.8.3 Type C Data

In type B experiments, we have the facility for making the pattern crosses overlap, which is typical of data arising from real problems. However, unlike type A, type B experiments do not allow us to define

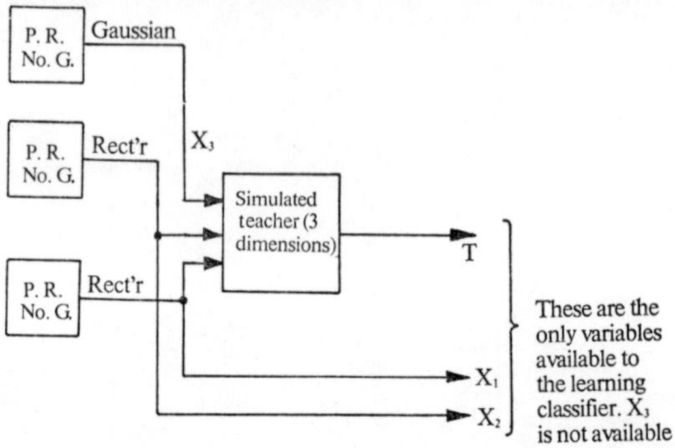

Figure 2.23 Method of generating type C synthetic data.

an ideal solution. It is possible to modify the method of generating type A data so that we can add a variable amount of overlap between classes, while still being able to define an ideal solution.

Consider Figure 2.23, which should be compared to Figure 2.20. We have added a pseudo-random number generator, X_3, which has a Gaussian probability density function. In the 2-space with axes

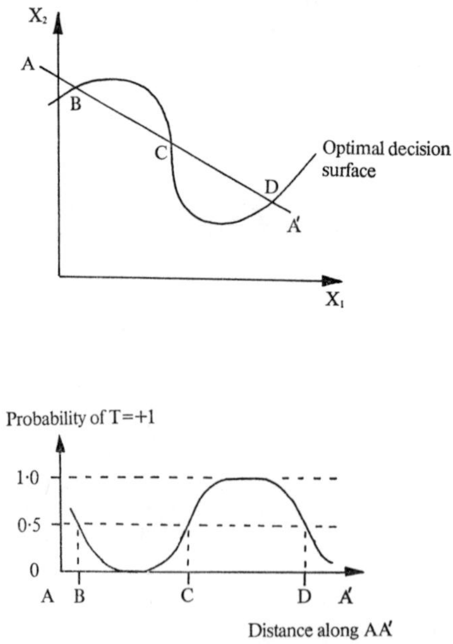

Figure 2.24 Type C synthetic data. (A) descriptor space showing the optimal decision surface. (B) section of the pdf along the line AA'.

(X_1, X_2), the teacher defines overlapping classes, even though they do not overlap in the 3-dimensional space, with axes (X_1, X_2, X_3).* The ideal decision surface in 2-space is the projection obtained by setting X_3 equal to zero in the 3-space. The teacher decision is indeterminate for points on the optimal surface, but becomes more definite as we move away from this surface (Figure 2.24).

The only objection to the use of type C data is that the probability distributions are essentially flat. This is untypical of real problems which probably have complex distributions, resembling mountain ranges with many peaks. Nevertheless types A and C data are useful for obtaining a comparison between the ideal and classifier decision surfaces. Type C data is particularly valuable for studying the effects of class overlap. We do not have scope in this short book for developing this topic. The generation and use of type B synthetic data. This topic is discussed at greater length in Appendix 3.

2.9 RELATION BETWEEN PATTERN RECOGNITION AND OTHER SUBJECTS

2.9.1 Design of Minimal Combinational Logic Circuits

Logic design is concerned with the construction of minimum-sized digital networks. The task required of the network is specified by a table of input-output relationships. Each output depends solely on the current state of the inputs. This table may therefore be regarded as a number of input vectors and their associated outputs. This format is, of course, identical with that of the pattern vector and its corresponding teacher decision.

The parallels between pattern classification and logic design may be developed even further. It is quite a common practice among logic circuit designers to use a hyperspace to represent the state of the inputs [18]. A space used in this way has a coarse quantisation, unlike the hyperspaces used for pattern description, which normally have a high resolution.†

The basic switching functions of logic design are relatively simple compared to those that we shall encounter in pattern classification (Figure 2.25). In both subjects, complex functions can be synthesised by combining these simpler basic functions.

The major difference between these two fields of study, is that logic design does not tolerate any errors, whereas it is inevitable that a pattern classifier will produce errors, if the classes overlap.

* We are simulating the destruction of information in the analyser by adding X_3 to the teacher.

† Coarsely quantised spaces are permissible in pattern classification, but are obligatory in logic design.

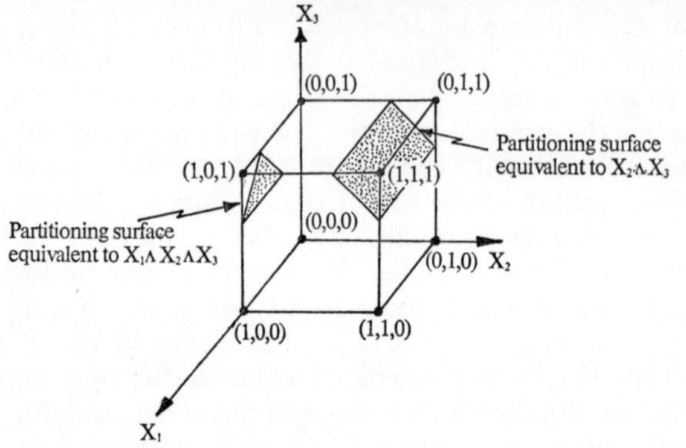

Figure 2.25 Drawing a parallel between switching functions in logic circuit theory and decision surfaces in pattern classification.

2.9.2 Adaptive Logic Circuits

The work of Aleksander [19] is important here, because it provides the logical equivalent of learning in a pattern classifier. The basis of Aleksander's work is a device called a SLAM (Stored Logic Adaptive Microcircuit), which is an integrated circuit RAM, with an input decoder (Figure 2.26). The contents of the RAM may be changed by

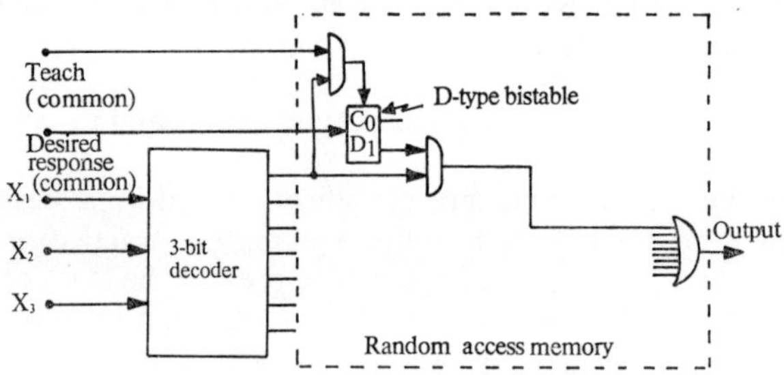

Figure 2.26 Circuit of SLAM-8. (Aleksander uses *R-S* flip-flops and gates, rather than D-type devices.)

applying a signal to the teach wire. Only that store with the address specified by the inputs (X_1, X_2, X_3) will be changed. By a suitable choice of inputs the RAM may be set to any one of 2^8 possible states. In fact, the SLAM-8* is a universal logic gate, since it can form any logic function of 3-variables. A universal logic gate of any size can be synthesised by connecting several SLAM-8's into a suitable network. To design the gate all we need to do is connect the inputs to the terminals X_1, X_2, . . . and indicate what response we require. The same

* SLAM-8 is the 3-input (8 store) version.

procedure is repeated for all inputs for which there is a specified output. This is all that is required to synthesise any switching function within the range of the SLAM network. This device is similar to a learning pattern classifier (Figure 2.27).

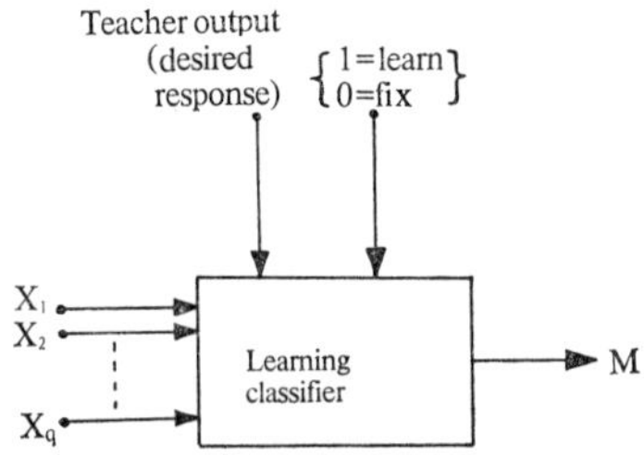

Figure 2.27 A SLAM and a learning classifier have the same kind of structure. The inputs to a SLAM are binary, whereas a classifier has analogue inputs.

2.9.3 Finite State Machines

Finite state machines are often said to *recognise* certain input sequences [20]. For example, the simple machine drawn in Figure 2.28 will enter state *D* after every occurrence of the input sequence (110). In this sense the machine does recognise sequences (patterns), although

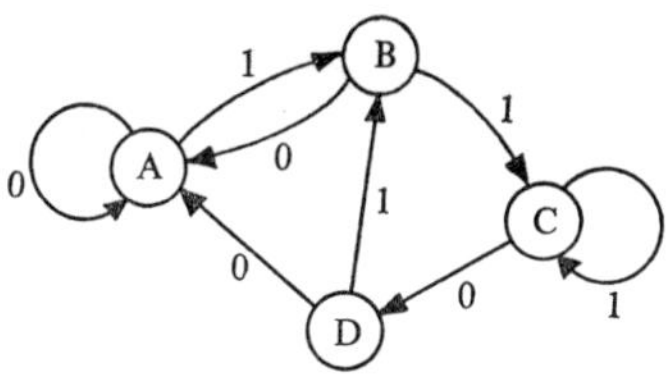

Figure 2.28 A finite state machine which recognises sequences ending in 110. If such a sequence is presented, the machine enters state D.

the type of recognition process is quite different from that found in a pattern classifier. The finite state machine is a *serial recogniser*, and is therefore more akin to sequential decision theory (section 2.9.8) than to the parallel decision process of a classifier.

2.9.4 Statistical Decision Theory

The conceptual foundations of our subject lie in statistics, although the techniques that we offer are unlikely to appeal to a conventional statistician. He normally prefers to work with procedures which have been established by formal analysis, whereas our work is pragmatic, and based on experimental evaluation. We see our work as existing in parallel with the more formal statistical methods. The author would be the first to welcome a rigorous mathematical background to his work, but *feels* that a pragmatic approach will yield results in a shorter time. There is another important feature in our work, which has no counterpart in statistics. It is important to ensure that the procedures we suggest can be easily realised in electronic hardware. We already possess a

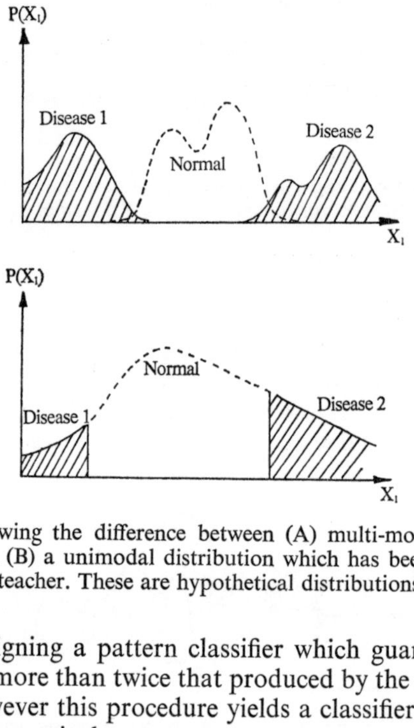

Figure 2.29 Showing the difference between (A) multi-modal probability distributions, and (B) a unimodal distribution which has been sectioned by the teacher. These are hypothetical distributions.

technique for designing a pattern classifier which guarantees that the error rate* is not more than twice that produced by the Bayes' classifier (section 5.4). However this procedure yields a classifier that is so large as to be totally impractical.

If we knew, or could describe, the probability density functions (pdfs), we could make use of the conventional Bayes' decision theory [2, 21, 22]. We cannot assume that we shall have any detailed knowledge about the pdfs since the definition of the pattern classes is so poorly understood. (We cannot even specify what constitutes a letter "A", nor what symptom-combinations are necessary for a diagnosis of migraine!) Since we cannot define the classes adequately we shall find it impossible to make any firm statements about the distribution of patterns in the **X**-space. However, another approach is feasible. It might be possible

* Proportion of disagreements between the classifier and teacher.

to produce a technique for describing the pdfs. A procedure of this kind has been suggested by Sebestyen [17], and has been refined further by Mucciardi and Gose [23]. This procedure, called adaptive sample set construction is discussed in Appendix 2. It attempts to model the pdfs, using the union of a number of Gaussian hills (Figure 2.22, sections 2.8 and 3.9). Once the model has been designed, the estimated pdfs can be used in Bayes' formula. Adaptive sample set construction illustrates the point we made earlier, that classification techniques are often suggested with no reference to their computation. (However, it would not be difficult, in this case, to produce an approximate method which could be more easily implemented.)

Another frequently encountered approach is to assume that the pattern classes have a Gaussian distribution [1, 2, 24]. (Sometimes other distributions are assumed, although far less frequently.) This is a rather sterile approach since we know that in many problems a class is multi-modal, or a section of a unimodal distribution* (see Figures 2.17 and 2.29). Certain classes do appear to be unimodal, although the assumption that there is a Gaussian distribution seems to be totally unjustified.

2.9.5 Numerical Taxonomy

Numerical taxonomy is a specialised branch of pattern recognition and is concerned with the classification of biological organisms into species, genera etc. Taxonomists frequently use hyperspaces to represent their "patterns", and measure similarity in terms of distance in such a space. They have become resigned to using small data sets [25], since they have been constrained to using conventional digital computers. Hence taxonomic methods are often incapable of being extended to handle the much larger sets which we shall encounter. Most taxonomic procedures have been developed around a matrix giving the inter-pattern distances. This matrix is square, and contains N^2 elements, where N is the number of patterns in the training set. A typical value for N might be 20 in a taxonomic problem, whereas in pattern classification, 10^3 would be more realistic. Hence we see the difficulty† of using taxonomic methods in pattern recognition, since the computation time generally rises as N^2.

One of the most valuable contributions of taxonomy has been the development of the dendrogram ([25], and Appendix 2 of this book). This is a graphic technique for displaying similarity relationships between patterns. It would be valuable to possess software for drawing dendrograms for really large data sets. To the author's knowledge this has not yet been developed. The importance of the dendrogram

* An example is encountered in medicine: the probability distribution of haemoglobin suggests that "anaemia" is the tail of a unimodal distribution, rather than a distinct cluster.

† A special-purpose system with parallel vector-processing capabilities could possibly overcome these objections.

is the facility it offers for human pattern recognition to be used in the difficult problem of counting and locating clusters.

2.9.6 Control Theory

A number of parallels may be drawn between pattern classification and control theory. The state-space approach, familiar to control engineers, is simply a hyperspace description of the system. The state of the system is represented by the position of a point within this space. The control engineer is expected to design a mechanism which will always return the plant to the optimal state, (this is represented by a goal region in the space). To do this, the controller will possibly adopt a number of distinct control strategies, depending upon the current position of the state-vector. The device for selecting the appropriate control strategy could be a pattern classifier. For example, Widrow and Smith [26] successfully trained a pattern classifier to control a bang-bang servo-mechanism for balancing an inverted pendulum.

2.9.7 Threshold Logic

A threshold logic gate [27] is a simple pattern classifier (Figure 2.30). (In the next chapter we shall see that it is a linear classifier.) The inputs

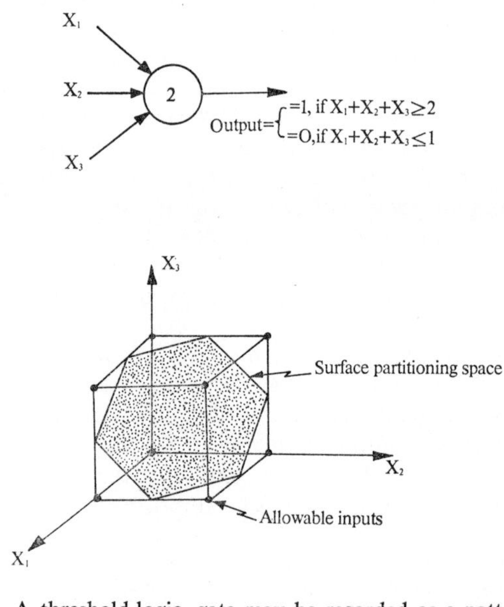

Figure 2.30 A threshold-logic gate may be regarded as a pattern classifier with binary inputs. (A) a gate with threshold equal to 2. (B) the decision surface equivalent to the switching function of a threshold-logic gate.

to a threshold gate are digital which is, of course, not always the case with a pattern classifier. For this reason, the techniques of threshold logic are unlikely to contribute significantly to pattern classification.

2.9.8 Sequential Decision Theory

A sequential decision tree may, but need not, operate on numeric data. Another feature distinguishes sequential decision theory from pattern classification. That is, the number of descriptors considered

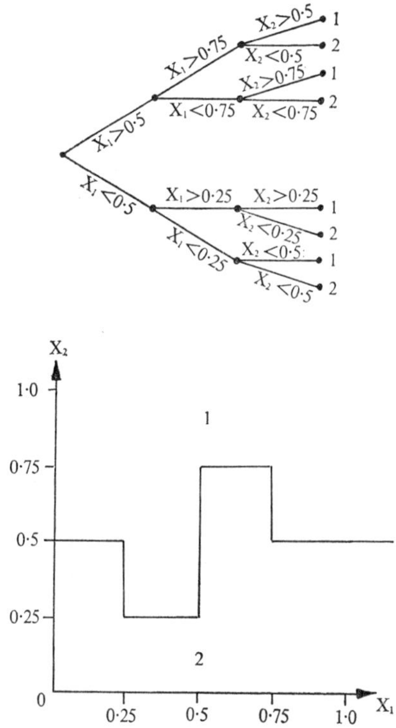

Figure 2.31 (A) a simple decision tree and (B) its equivalent decision surface.

before reaching a decision may vary at different parts of the tree. However, it is possible to draw a parallel between these two fields. For example, we can always draw a surface which yields the same decision as a tree, if the latter operates only on numeric data (Figure 2.31). However, it may not be so easy to define a tree which approximates a given decision surface.

One significant advantage of a tree is found when some of the descriptors are missing. It is still possible to calculate a decision, whereas our present model for pattern classification does not allow this.

REFERENCES

1. H. C. Andrews, Introduction to Mathematical Techniques in Pattern Recognition (Wiley, New York, 1972).
2. E. A. Patrick, Fundamentals of Pattern Recognition (Prentice-Hall, Englewood Cliffs, N.J., 1972).
3. B. R. Wilkins and B. G. Batchelor, Evolution of a Descriptor Set for Pattern Recognition, *Proc. Conference on Technical and Biological Problems of Control, Instrument Society of America*, 794–9, 1970.
4. A. N. Mucciardi and E. E. Gose, Evolutionary Pattern Recognition in Incomplete Non-linear Multi-threshold Networks, *Trans. I.E.E.E., EC-15*, **2**, 257–262, 1966.
5. D. M. Y. Somerville, Introduction to the Geometry of *N*-Dimensions (Methuen, London, 1929).
6. G. C. Shephard, Vector Spaces of Finite Dimension (Oliver and Boyd, Edinburgh, 1966).
7. R. A. Fisher, The use of Multiple Measurements in Taxonomic Problems, *Annals of Eugenics*, **7**, 179, 1936.
8. D. M. Balston, Measurement Systems for Visual Pattern Recognition, *Ph.D. Thesis*, Southampton, 1972.
9. P. Fourman, Calcium Metabolism and the Bone (Blackwell, Oxford, 1960).
10. D. N. Baron, Short Textbook of Chemical Pathology (English Universities' Press, London, 1970).
11. J. C. Houston, C. L. Joiner and J. R. Trounce, A Short Textbook of Medicine (English Universities' Press, London, 1968).
12. S. H. Lavington, Computer Simulation of a Speech Recognition System, *Proc. I.E.E.*, **116**, 6, 1053–9, 1969.
13. R. P. Chambers, Random Number Generation on Digital Computers, *I.E.E.E., Spectrum*, **4**, 48–56, 1967.
14. T. G. Newman and P. L. Odell, Generation of Random Variates (Griffin, London, 1971).
15. B. G. Batchelor, Automatic procedures for varying the complexity of a pattern classifier, in Machine perception of patterns and pictures, 317–327, *Institute of Physics, Conference Series*, No. 13, 1972.
16. B. G. Batchelor, Growing and Pruning a Pattern Classifier, *Information Sciences*, **6**, 97–122, 1973.
17. G. S. Sebestyen, Decision-making Processes in Pattern Recognition (Macmillan, New York, 1962).
18. F. J. Hill and G. R. Peterson, Introduction to Switching Theory and Logical Design (Wiley, New York, 1968).
19. I. Aleksander, Microcircuit Learning Computers (Mills and Boon, London, 1971).
20. F. C. Hennie, Finite-state Models for Logical Machines (Wiley, New York, 1968).
21. D. Middleton, An Introduction to Statistical Communication Theory (McGraw-Hill, London, 1960).
22. R. von Mises, Mathematical Theory of Probability and Statistics (Academic Press, New York, 1964).
23. A. N. Mucciardi and E. E. Gose, An Automatic Clustering Algorithm and its Properties in High-Dimensional Spaces, *Trans. I.E.E.E., SMC-2*, 2, 247–254, 1972.
24. N. J. Nilsson, Learning Machines (McGraw-Hill, New York, 1965).
25. R. R. Sokal and P. H. A. Sneath, Principles of Numerical Taxonomy (Freeman, San Francisco, 1963).
26. B. Widrow and F. W. Smith, Pattern Recognising Control Systems, in Computer and Information Sciences, 288–317 (Spartan, Washington, 1964).
27. S. L. Hurst, Threshold Logic (Mills and Boon, London, 1972).

Chapter 3

Methods of Pattern Classification

3.1 INTRODUCTION

One of the most bewildering aspects of our subject is the large number of schemes that have been suggested for classification. At first sight, it seems that there have been as many suggestions as authors. When we look closer we see that there are four or five basic techniques for classification, and that the remainder are merely refinements of these. So far, there has been little attempt to compare the various methods of classification. This is probably due to their obvious differences, which have, for a long time, hidden their similarities. In the following chapter, we shall attempt to show that most of the classifiers described in the literature are simply variations upon the same basic theme. Once this is appreciated it will become easier to decide which classifiers are best. This chapter is a catalogue of the most significant classifiers, and prepares us for this comparison.

3.2 LINEAR CLASSIFIER

The linear classifier [1, 2, 3] is one of the simplest pattern classifiers, yet it is still frequently encountered in the literature, despite the recent discovery of other, more powerful techniques. The major reasons for this continuing popularity are that it is well understood, it can readily be implemented in electronic hardware, and training methods exist which have been proved to converge. It should also be added that many problems can be solved quite adequately using a linear classifier. (In such a situation it would generally be unjustified to use a more complex technique.)

In a 2-dimensional space the decision surface of a linear classifier is a straight line, while in 3 dimensions we obtain a plane, and in q dimensions a hyperplane. The orientation of this hyperplane can be determined by q parameters, which we shall call weights, and shall denote by $W_1, \ldots, W_q$. One further parameter is required, which will be represented by V. The decision (M) of a two-class linear classifier is calculated as follows:

$$M = \operatorname{sgn}\left(V + \sum_{i=1}^{q} W_i X_i\right) \tag{3.1}$$

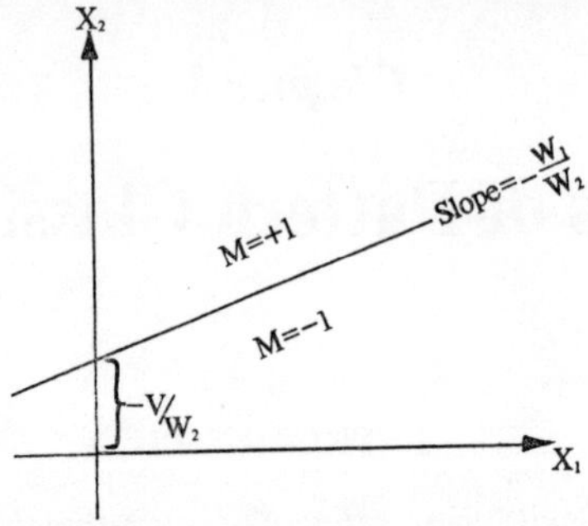

Figure 3.1 Decision surface of a linear classifier in a 2-dimensional space.

$M = +1$ on one side of the hyperplane, and -1 on the other (Figures 3.1 and 3.2). By varying V we shift the hyperplane to a new position parallel to the old one. If V is zero the hyperplane intersects the origin. In later chapters, we shall find that it is more convenient to use the following homogenous form for the linear classifier:

$$M = \operatorname{sgn}\left(\sum_{i=1}^{q+1} W_i X_i\right) \tag{3.2}$$

where

$$\left.\begin{aligned} W_{q+1} &\equiv V \\ X_{q+1} &\equiv 1 \end{aligned}\right\} \tag{3.3}$$

3.3 ϕ-MACHINE

To achieve a more flexible decision surface we could add a descriptor transformation network to the linear classifier (Figure 3.3). If the

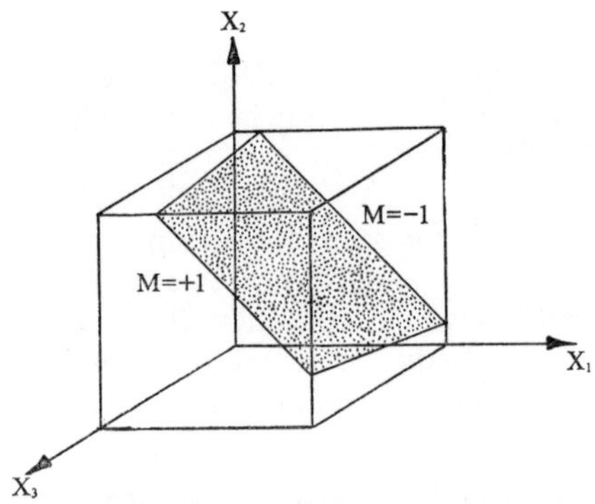

Figure 3.2 Decision surface of a linear classifier in a 3-dimensional space.

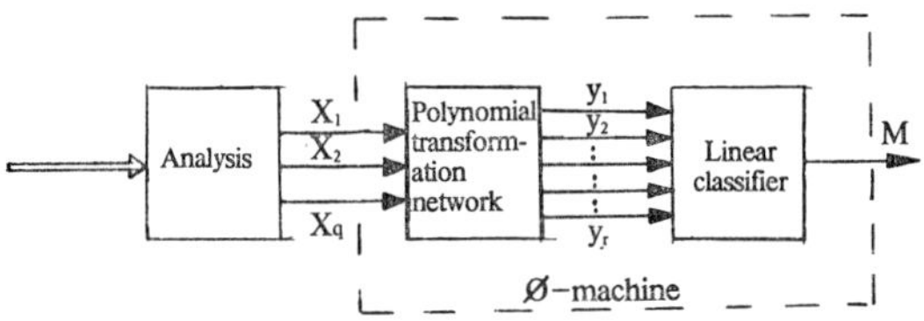

Figure 3.3 Structure of a ϕ-machine.

transformation network calculates polynomial terms the resulting classifier is called a ϕ-machine [3].

A simple ϕ-machine might calculate the following:

$$\left.\begin{aligned} Y_1 &= X_1 \\ Y_2 &= X_2 \\ Y_3 &= X_1^2 \\ Y_4 &= X_1^3 \\ Y_5 &= 1 \\ M &= \operatorname{sgn}\left(\sum_{i=1}^{5} W_i Y_i\right) \end{aligned}\right\} \tag{3.4}$$

A typical decision surface from a ϕ-machine is shown in Figure 3.4. Notice, however, that the decision surface of a ϕ-machine need not be continuous.

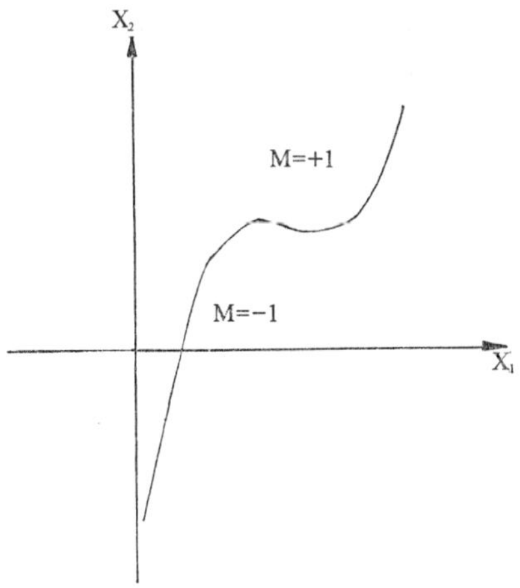

Figure 3.4 Typical decision surface produced by a ϕ-machine in a 2-dimensional space.

It should be also noted that the W_i are contained in a linear equation. (This allows us to use the same learning rules for a ϕ-machine and linear classifier.)

3.4 PIECE-WISE LINEAR CLASSIFIER

The piece-wise linear classifiers (PWLC) are capable of realising complex decision surfaces without using the expensive polynomial transformation network required by a ϕ-machine. Many forms of PWLC have been suggested, but they all consist of at least two layers of units.* The bottom layer contains a number (N_1) of linear sub-classifiers. The higher layers combine the decisions from the bottom layer (Figure 3.5). The combinational logic may be implemented using

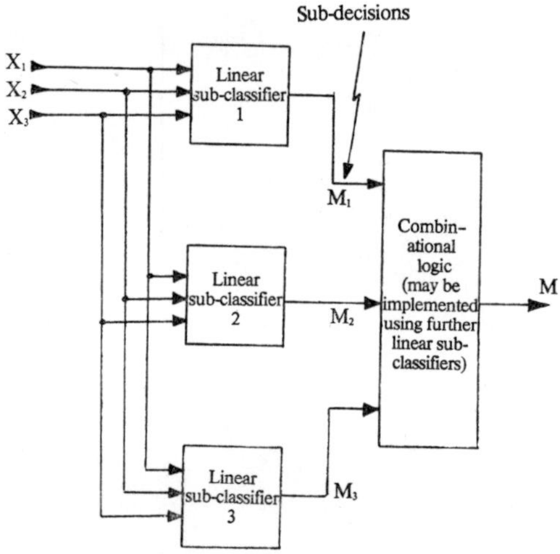

Figure 3.5 Structure of a piece-wise linear classifier.

conventional digital logic, or a layered network of linear subclassifiers. The latter alternative has the advantage that its logical function may be modified by varying the weights in the second and higher layers.

Figures 3.6 and 3.7 show decision surfaces produced by a PWLC in 2 and 3-dimensional spaces. Any surface that can be represented by a finite number of hyperplanar segments can be realised by a PWLC.

* To avoid ambiguity, we shall use the term subclassifier to indicate a classifier-like device which forms part of a larger system, for which we shall reserve the name classifier.

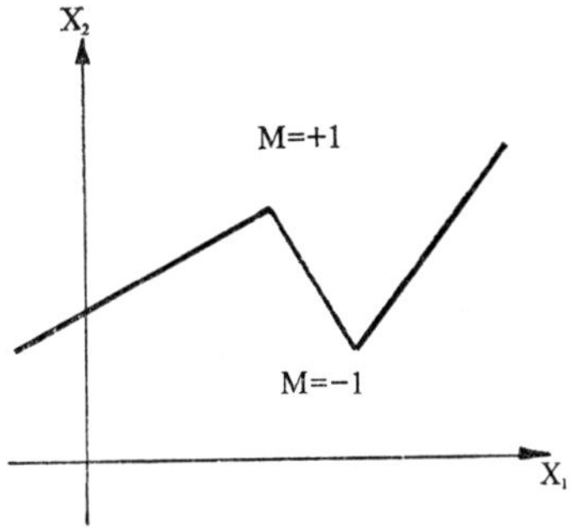

Figure 3.6 Decision surface of a piece-wise linear classifier in a 2-dimensional space.

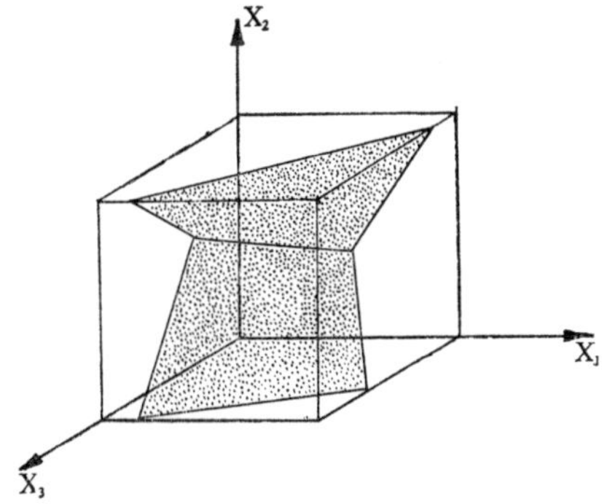

Figure 3.7 Decision surface of a piece-wise linear classifier in a 3-dimensional space.

3.5 NEAREST NEIGHBOUR CLASSIFIER

The three classification techniques described so far are all based on the linear classifier. We are now going to depart from this approach. Superficially there is little resemblance between the Nearest Neighbour Classifier (NNC) and the remaining schemes to be discussed in this chapter. However, it will soon become apparent that they store information in the same way. We begin by describing the NNC which is possibly the simplest scheme using this storage organisation.

The NNC makes use of the correspondence between similarity and distance, discussed in the previous chapter. Imagine a machine which stores a number of representative **X**-vectors from each class. Each new

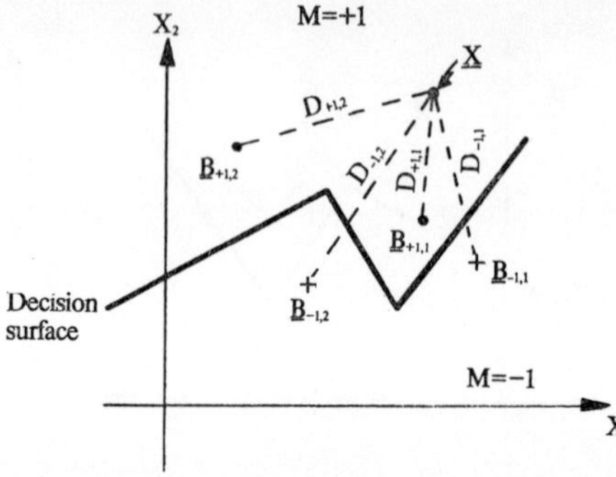

Figure 3.8 Decision surface of a Nearest Neighbour Classifier in a 2-dimensional space. Notice that this surface is identical to that shown in Figure 3.6. To find the classification of the point **X**, we calculate the distances $D_{-1,1}$, $D_{-1,2}$, $D_{+1,1}$, $D_{+1,2}$, and find the minimum (in this case $D_{+1,1}$). The classification for this point is $+1$.

pattern is then assigned to the class of the most similar stored representative. If we assume that there is an inverse relationship between similarity and distance in the descriptor space, we must minimise the distance between the **X**-vector and the stored representatives. This machine is the NNC.

The storage in an NNC is divided into areas in which we can store one q-dimensional vector, called a *locate*. Let there be:

(a) N_{-1} locates, denoted by $\mathbf{B}_{-1,1}, \ldots, \mathbf{B}_{-1,N_{-1}}$. These locates will be associated with the NNC decision, $M = -1$.
(b) N_{+1} locates, denoted by $\mathbf{B}_{+1,1}, \ldots, \mathbf{B}_{+1,N_{+1}}$. These locates will be associated with the NNC decision, $M = +1$.

In addition we shall let:

$$\mathbf{B}_{i,j} = (B_{i,j,1}, \ldots, B_{i,j,q}) \tag{3.5}$$

As usual, q is the dimensionality of the space.

Let $D_{i,j}$ be the Euclidean distance between the locate $\mathbf{B}_{i,j}$ and the **X**-vector. Then

$$D_{i,j} = \sqrt{\sum_{k=1}^{q} (B_{i,j,k} - X_k)^2} \tag{3.6}$$

The decision of the NNC is given by M where

$$M = \operatorname{sgn}\left(\min_{j=1,\ldots,N_{-1}} (D_{-1,j}) - \min_{j=1,\ldots,N_{+1}} (D_{+1,j}) \right) \tag{3.7}$$

In the simple example shown in Figure 3.8 we find that

$$D_{+1,1} < D_{-1,1} < D_{+1,2} < D_{-1,2}$$

The decision surface of a NNC is piece-wise linear. In fact it can realise any such surface, if there are enough locates. To emphasise this point, we have deliberately made the decision surfaces in Figures 3.6 and 3.8 identical.

We do not wish to give the impression that the locates are necessarily representatives of the pattern classes. As far as we are concerned, the locates are simply points in X-space which help us to draw the decision surface. They will be particularly useful in later chapters when we consider the way the decision surface changes during learning.

3.6 VARIATIONS OF THE NNC

3.6.1 Alternative Distance Measures

In the preceding section, we defined the NNC in terms of the Euclidean distance. Since this is relatively difficult to calculate in analogue hardware, various other distances [4] have been proposed:

(i) The city block distance (sometimes called Manhatten distance):

$$\sum_{K=1}^{q} |B_{I,J,K} - X_K| \tag{3.8}$$

(ii) The square distance:

$$\max_{K=1,\ldots,q} \quad |B_{I,J,K} - X_K| \tag{3.9}$$

(iii) An octagonal distance [5], equal to a weighted sum of the city block and square distances:

$$\sum_{K=1}^{q} |B_{I,J,K} - X_K| + \sqrt{q} \max_{K=1,\ldots,q} \quad |B_{I,J,K} - X_K| \tag{3.10}$$

The advantage of these distances lies in their computational simplicity, but as we shall see later they introduce some strange effects. It has been proposed that any of these might be used in lieu of the Euclidean distance in the NNC. Figures 3.9–3.11 compare a simple NNC using the Euclidean distance to others which use the same (two) locates but other distances.

3.6.2 Adding Weights to the NNC

Our last modification to the NNC is the introduction of weights. We shall denote these by $W_{-1,1}, \ldots, W_{-1,N_{-1}}$ and $W_{+1,1}, \ldots, W_{+1,N_{+1}}$.

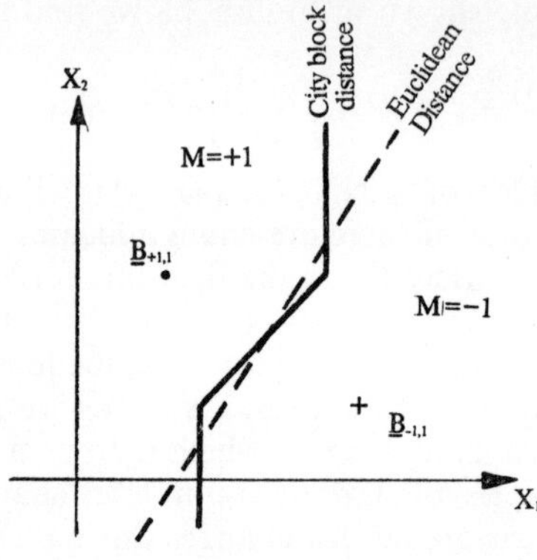

Figure 3.9 Comparison of the decision surfaces of two simple NNCs which use the same two locates, but different distance measures (Euclidean and city block distances).

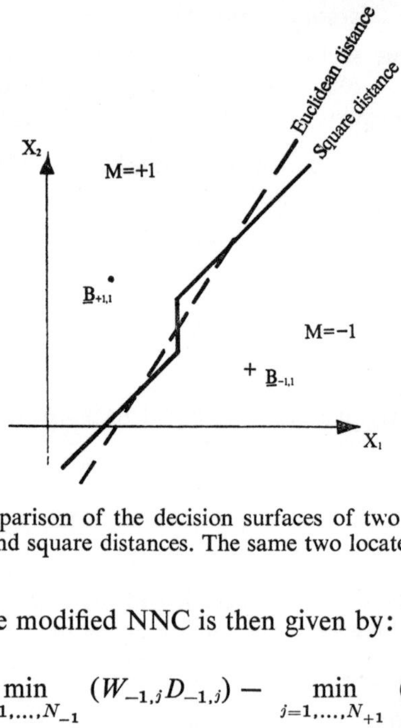

Figure 3.10 Comparison of the decision surfaces of two NNCs using the Euclidean and square distances. The same two locates are used.

The decision of the modified NNC is then given by:

$$M = \text{sgn}\left(\min_{j=1,\ldots,N_{-1}} (W_{-1,j}D_{-1,j}) - \min_{j=1,\ldots,N_{+1}} (W_{+1,j}D_{+1,j})\right) \tag{3.11}$$

Of course the $D_{i,j}$ could be calculated using any of the distances defined in expressions (3.6, 3.8, 3.9 or 3.10). If the Euclidean distance is used, the decision surface consists of fragments of hyperspheres, and is thus said to be piece-wise hyperspherical (Figure 3.12). The classifier defined

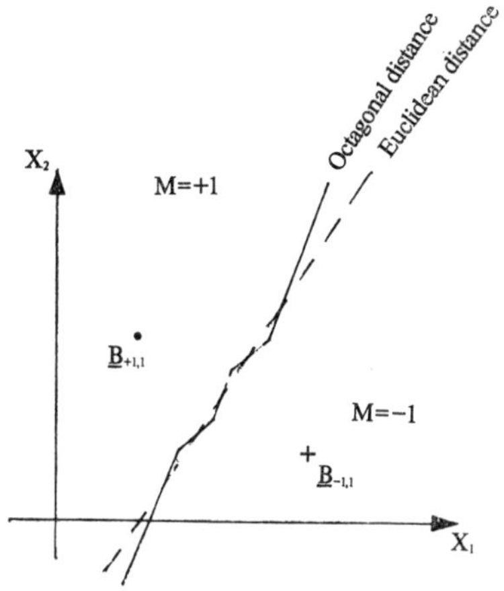

Figure 3.11 Comparison of the decision surfaces of two NNCs using the Euclidean and octagonal distances. The same two locates are used.

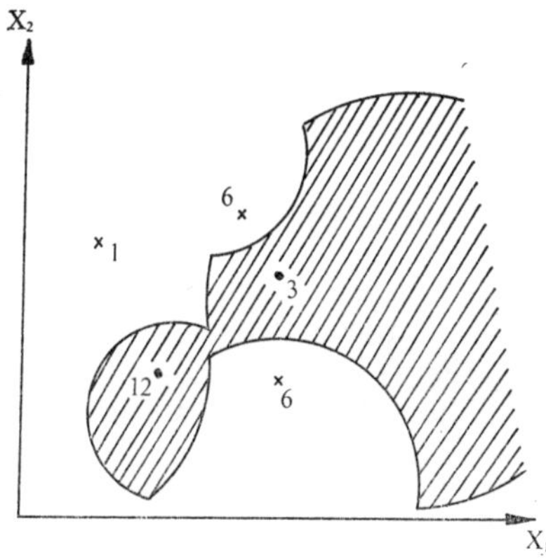

Figure 3.12 Decision surface of a modified NNC in a 2-dimensional space. Numbers denote values of the weights, W_i.

by equation (3.11) will be called the *modified NNC* to distinguish it from the normal form (equation (3.7)).

3.7 POTENTIAL FUNCTION CLASSIFIER

Potential Function Classifier (PFC) is a term applied to a large number of pattern classifiers, which can be defined by the same form of equation. Like the modified NNC, the PFCs employ both locates and weights. We shall therefore continue to use the notation developed in the two previous sections.

A typical PFC might calculate a decision M using the following rule:

$$M = \operatorname{sgn}\left(\sum_{j=1}^{N_{+1}}\left(\frac{W_{+1,j}}{D_{+1,j}}\right) - \sum_{j=1}^{N_{-1}}\left(\frac{W_{-1,j}}{D_{-1,j}}\right)\right) \tag{3.12}$$

where $D_{i,j}$ is given by equation (3.6). This is just one example from a range of classifiers, which can all be defined by the following form of equation:

$$M = \operatorname{sgn}\left(\sum_{j=1}^{N_{+1}} P(D_{+1,j})W_{+1,j} - \sum_{j=1}^{N_{-1}} P(D_{-1,j})W_{-1,j} + T\right) \tag{3.13}$$

where T is a scalar quantity which can be zero, positive or negative, and $P(D)$ is a monotonically decreasing function of D. Equation (3.13) defines the general two-class PFC.

The name Potential Function Classifier was suggested by the Russian workers, who first proposed this method of classification [6, 7]. They imagined that the q-dimensional descriptor space contained negative electrical charges of magnitude $W_{-1,1}, \ldots, W_{-1,N_{-1}}$ placed at $\mathbf{B}_{-1,1}, \ldots, \mathbf{B}_{-1,N_{-1}}$; similarly for positive charges $W_{+1,1}, \ldots, W_{+1,N_{+1}}$ at $\mathbf{B}_{+1,1}, \ldots, \mathbf{B}_{+1,N_{+1}}$. They then considered the electro-static potential, and noted that the sign of the potential at a point $\mathbf{X}$ is given by (3.12). The zero-potential surface in the electrostatic model is identical with the decision surface of the classifier defined by equation (3.12).

3.8 VARIATIONS OF THE PFC

The PFC was first envisaged through the use of the electro-static analogue, which is now of little more than historic interest. Other forms of potential decay function ($P(D)$ in (3.13)) have been proposed [8–10], often to avoid the infinite values which can occur in (3.12) (when the $\mathbf{X}$-vector is coincident with any of the locates). Another suggestion makes use of the following potential decay function

$$P(D) = 1/D^2 \tag{3.14}$$

This would allow us to eliminate the square-root operation required by (3.6). Many other possibilities exist. For example, we could use any of the distances defined in (3.8) to (3.10), rather than the Euclidean distance.

There are even more variations that we must consider. One of the most significant is the unipolar PFC [8], which has an equation of the form:

$$M = \operatorname{sgn}\left(T - \sum_{j=1}^{N_{-1}} P(D_{-1,j})W_{-1,j}\right) \tag{3.15}$$

This type of classifier stores locates for only one class. Equation (3.15) is a special case of (3.13), with $N_{+1} = 0$. The author prefers the term *homopolar* rather than *unipolar*. We can then use the term *heteropolar*

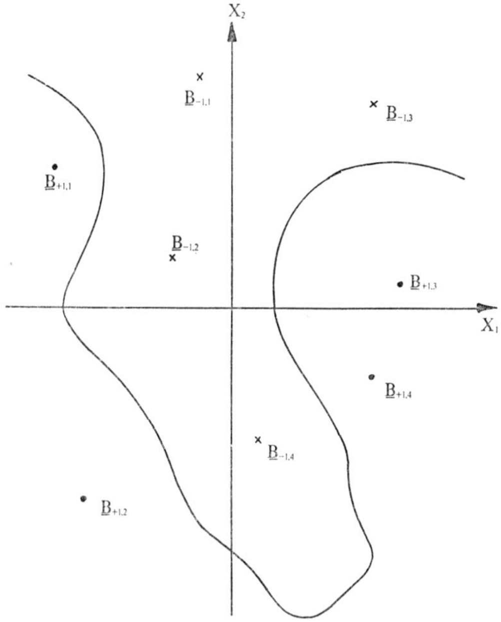

Figure 3.13 Decision surface of a heteropolar Potential Function Classifier in a 2-dimensional space.

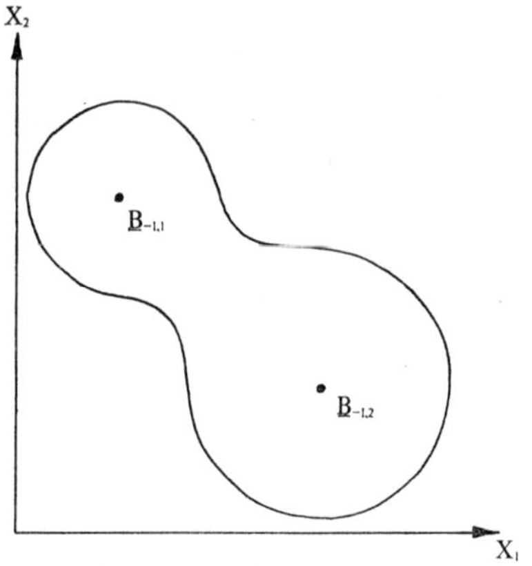

Figure 3.14 Decision surface of a homopolar Potential Function Classifier in a 2-dimensional space.

for those PFCs which store locates for each class. (Equation (3.12) defines a typical heteropolar PFC.)

With such an enormous range of possibilities, it is difficult to choose typical decision surfaces to illustrate the power of the PFCs. We have attempted to do this in Figures 3.13 and 3.14 for both the heteropolar and homopolar PFCs.

3.9 CLASSIFIER DESCRIBED BY G. S. SEBESTYEN

In 1962 G. S. Sebestyen published his book [1] which has now become a standard text for pattern recognition workers. In this book he describes a technique for classification which closely resembles the PFCs although it preceded them by about 3 years. Like the heteropolar PFC, the Sebestyen classifier (SC) calculates its decision using equation (3.13) in which:

$$\left.\begin{aligned} T &= 0 \\ P(D) &= \exp(-\alpha D^2) \qquad \text{where } \alpha \text{ is a positive scalar} \end{aligned}\right\} \tag{3.16}$$

3.10 COMPOUND CLASSIFIER

Imagine a handful of coins have been dropped on to a table. The "shadow" they cast is a typical decision surface produced by a Compound Classifier (Figure 3.15). In a 3-dimensional space the decision

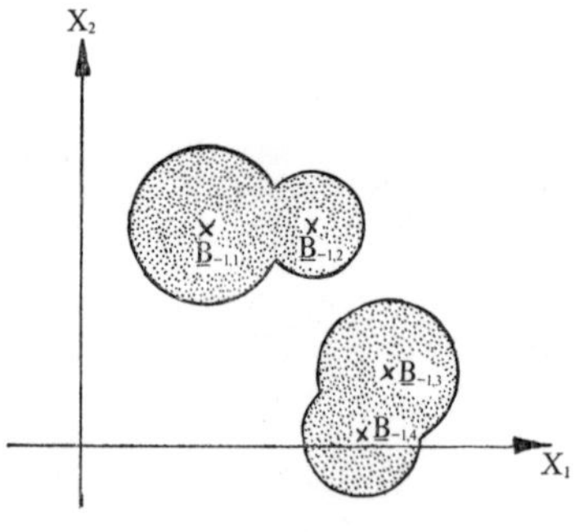

Figure 3.15 Decision surface of a Compound Classifier in a 2-dimensional space.

surface consists of a number of spheres which may, or may not, overlap. In multi-dimensional spaces the decision surface is piece-wise hyperspherical.

The Compound Classifier (CC) was first conceived [12, 13] as a layered device (Figure 3.16):

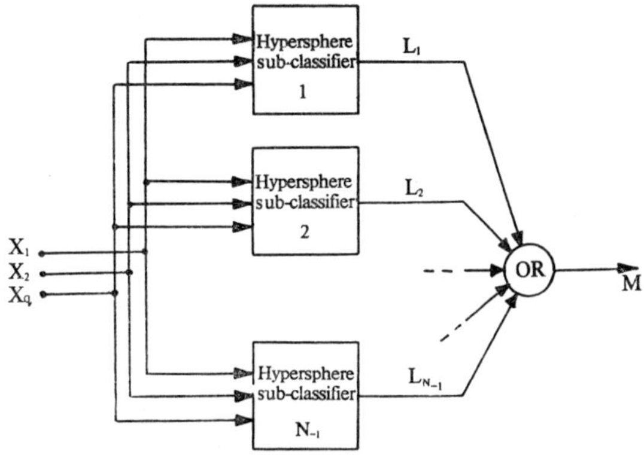

Figure 3.16 Structure of a Compound Classifier.

(a) the bottom layer contains a number of subclassifiers. Each of these decides whether the input point **X** lies inside a hypersphere. Such a device will be termed a hypersphere subclassifier.
(b) the top layer combines the sub-decisions from the bottom layer by forming their logical union.

Each subclassifier requires the storage of one q-dimensional locate to define the position of the hypersphere. Its size can be fixed by one further parameter. It will be convenient in the next chapter to have a description of the CC in terms of the double suffix notation already developed for the NNC and PFC. We shall assume then that there are N_{-1} locates: $\mathbf{B}_{-1,1}, \ldots, \mathbf{B}_{-1,N_{-1}}$. The size parameters will be represented by $F_1, \ldots, F_{N_{-1}}$. The jth sub-decision U_j is calculated as follows:

$$U_j = \text{sgn}\,(F_j - D_{-1,j}) \tag{3.17}$$

where $D_{-1,j}$ is defined by equation (3.6). The overall decision of the CC (M) is then calculated as follows:

$$M = \begin{cases} +1, \bigcup\limits_{j=1}^{N_{-1}} (U_j = +1) \\ -1, \text{ otherwise} \end{cases} \tag{3.18}$$

Equations (3.17) and (3.18) are the original form of definition of the CC. In order to compare the CC to the other classifiers we shall find it more convenient to rewrite the definition in terms of the minimum selector function:

$$M = \text{sgn}\left(1 - \min_{j=1,\ldots,N_{-1}} (D_{-1,j}/F_j)\right) \tag{3.19}$$

It is quite easy to verify that equations (3.18) and (3.19) give identical values for M. Equation (3.18) reflects the two-layer structure shown in

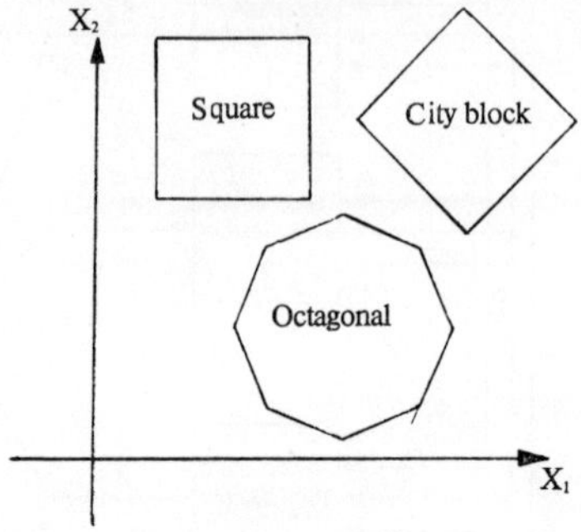

Figure 3.17 Alternatives to the circle obtained by using the non-Euclidean distances.

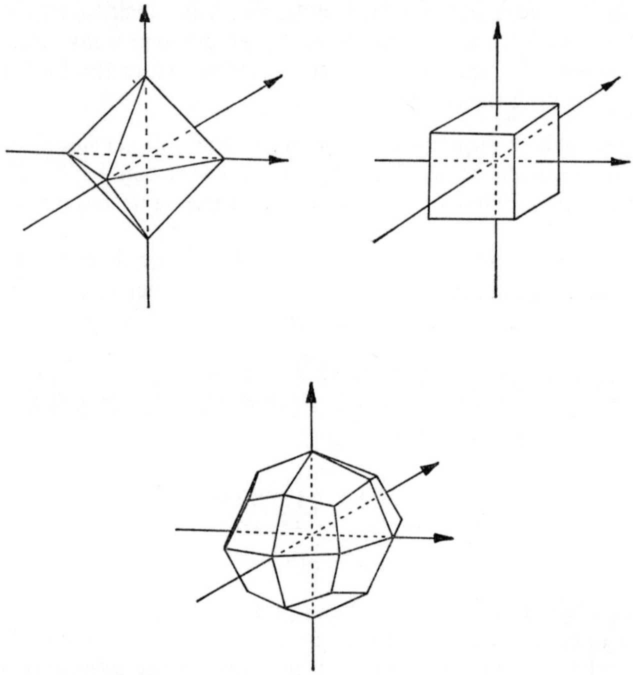

Figure 3.18 Alternatives to the sphere, obtained by using the non-Euclidean distances:

(A) city block distance
(B) square distance
(C) octagonal distance.

(Reproduced by permission of the Institution of Electrical Engineers)

Figure 3.15, while (3.19) will allow us to compare the CC with the other classifiers.

By now the reader will not be surprised that the CC can use the alternative distances defined in (3.8)–(3.10). The decision surfaces then consist of elements like those shown in Figures 3.17 and 3.18.

3.11 THE *k*-NEAREST NEIGHBOUR CLASSIFIER

The description of the *k*-Nearest Neighbour Classifier is only included here for completeness. The author feels that it is not particularly useful* since its decision surface can be duplicated by other simpler classifiers.

The first point that should be made is that the ordinary NNC (section 3.5) is a special case of the *k*-NNC with *k* equal to one. The *k*-NNC uses locates, but not weights.† Let the locates again be denoted by $\mathbf{B}_{i,j}$ ($i = \pm 1, j = 1, \ldots, N_i$). Let the distances‡ from the locate $\mathbf{B}_{i,j}$ to the input pattern $\mathbf{X}$ be $D_{i,j}$. The *k*-NNC selects those locates whose rank-distances from $\mathbf{X}$ are less than or equal to *k*. The decision of the *k*-NNC is then found by choosing the more frequently occurring class, among this group of *k* closest locates.

The decision surface of the *k*-NNC‡ is piece-wise linear and can therefore be exactly modelled by the ordinary NNC or the PWLC. For this reason, we feel that there is little point in continuing to discuss the *k*-NNC in any great detail. The reader with a particular interest in this topic will find a more complete description in [2].

3.12 SUMMARY

This chapter has listed over 20 classifiers ranging from the simple linear classifier to the more complex machines, such as the PFC. No

* It does have an intuitive appeal for many pattern recognition workers, which accounts for its continuing popularity [2]. Our approach assumes that it is impossible to store more than a few locates (perhaps 20). Other workers are prepared to accept a general-purpose digital computer as a solution to the pattern recognition problem (we are not), and are therefore prepared to store either the complete training set, or most of it. They make little or no attempt to eliminate atypical locates, and would find that an ordinary NNC produces many small "islands" of errors in the measurement space. These "islands" may be eliminated by employing the *k*-NNC, which forms a "consensus opinion" from the *k* closest locates, thereby achieving a rather smoother decision surface than the ordinary NNC using the same locates (Figure 3.19).

† Of course, it is possible to suggest a complicated version of the *k*-NNC which does use weights. There is little merit apparent.

‡ Again we have the option of using the Euclidean, city block, square or octagonal distances. The *k*-NNC has a piece-wise linear decision surface when using the Euclidean distance.

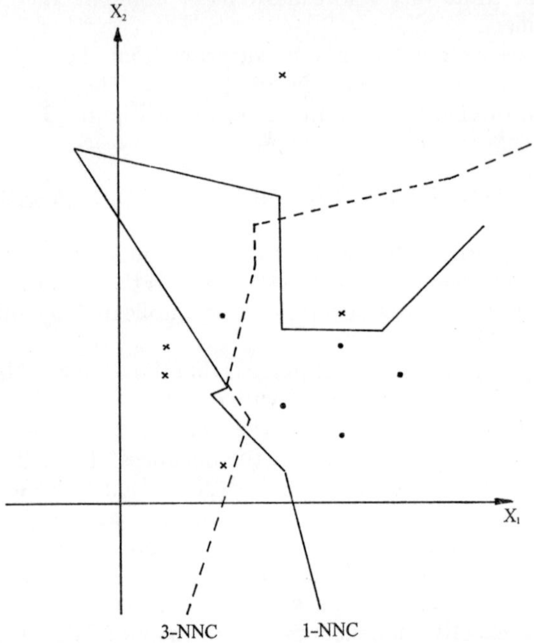

Figure 3.19 Comparison of the decision surfaces of the NNC (ordinary NNC, or 1-NNC) and the *k*-NNC. These classifiers use the same locates. Notice that the 3-NNC (and, in general, the *k*-NNC) produces less complicated surfaces, having fewer "corners". Hence the 1-NNC is capable of greater efficiency, although its locates must be carefully selected, since the effect of a single locate is more pronounced. In this example the surface due to the 3-NNC could be realised by a 1-NNC with only 7 locates.

Table 3.1 Summary of the principal methods of pattern classification (it is assumed that Euclidean distances are used)

Classifier	Equation	Decision surface
Linear	3.2	Hyperplane
ϕ-machine	—	Polynomial
Piece-wise linear (PWLC)	—	Fragments of hyperplanes
Nearest Neighbour (NNC)	3.7	Piece-wise linear
Modified NNC	3.11	General piece-wise hypersphere
Potential Function (PFC)	3.13	Not simply described
Homopolar PFC	3.15	Not simply described
Sebestyen's Classifier (SC)	3.16	Variant of PFC
Compound Classifier (CC)	3.17–18 or 3.19	Union of hyperspheres
k-NNC	—	Piece-wise linear

Table 3.2 Summary of the four distance functions

Distance	Equation	Equivalent of circle	Values compared to Euclidean distance: Minimum	Maximum
Euclidean	3.6	Circle	1.00	1.00
Square	3.8	Square (sides parallel to axes)	$\frac{1}{\sqrt{q}}$	1.00
City block	3.9	Square (sides 45° to axes)	1.00	$\sqrt{q}$
Octagonal	3.10	Regular octagon	$\frac{2+\sqrt{q}}{\sqrt{2}}$	$\sqrt{(\sqrt{q}+q)2}$

doubt the newcomer to the subject is wondering why we have so many different schemes. The reason is that nobody has yet simplified the subject by indicating which classifiers are worth studying further, and which should be discarded as useless. To summarise this chapter, we have listed the most significant classifiers in Table 3.1. The properties of the four alternative distances are summarised in Table 3.2.

REFERENCES

1. H. C. Andrews, Introduction to Mathematical Techniques in Pattern Recognition (Wiley, New York, 1972).
2. E. A. Patrick, Fundamentals of Pattern Recognition (Prentice-Hall, Englewood Cliffs, N.J., 1972).
3. N. J. Nilsson, Learning Machines (McGraw-Hill, New York, 1965).
4. A. Rosenfeld and J. L. Pfaltz, Distance Functions on Digital Pictures, *Pattern Recognition*, **1**, 33–61, 1968.
5. B. G. Batchelor, Improved Distance Measure for Pattern Recognition, *Electronics Letters*, **7**, 18, 521–4, 1971.
6. M. A. Aiserman, E. M. Braverman and L. I. Rozonoer, Potential Functions Technique and Extrapolation in Learning System Theory, *Proc. Congr., I.F.A.C.*, 3rd June, London, 1966.
7. A. G. Arkadev and E. M. Braverman, Teaching Computers to Recognise Patterns, (Academic Press, London, 1967).
8. C. J. W. Mason and C. H. McFall, Some Experiences with the Method of Potential Functions, in *Pattern Recognition, I.E.E. Conf. Publication*, **42**, 69–76, 1968.
9. B. G. Batchelor, N. L. Ford and B. R. Wilkins, Family of Pattern Classifiers, *Electronics Letters*, **6**, 12, 386–9, 1970.
10. B. G. Batchelor, N. L. Ford and B. R. Wilkins, Learning in a Potential Function Classifier, *Electronics Letters*, **6**, 25, 826–8, 1970.
11. G. S. Sebestyen, Decision-making Processes in Pattern Recognition (Macmillan, New York, 1962).
12. B. G. Batchelor, Learning Machines for Pattern Recognition, *Ph.D. Thesis* Southampton, 1969.
13. B. G. Batchelor and B. R. Wilkins, Adaptive Discriminant Functions, in *Pattern Recognition, I.E.E. Conf. Publication*, **42**, 168–178, 1968.

Chapter 4

Comparison of Classifiers

4.1 INTRODUCTION

In the preceding chapter we described or hinted at the existence of over 20 classifiers. The reader will probably feel that there is little unity in a subject which has so many techniques available, with no indication as to which are the most useful. This lack of a coherent theory has done much to discredit pattern recognition in recent years.

This chapter is an attempt to provide such a unifying theory. The major points of our theory may be summarised as follows:

(a) The CC is a homopolar version of the modified NNC.
(b) The decision surfaces of the modified NNC and heteropolar PFC are similar.
(c) The decision surfaces of the CC and the homopolar PFC are similar.
(d) The decision surfaces of certain classifiers are extremely sensitive to small changes in the stored parameters.

By the time we reach the end of this chapter we should be in a position to choose the best classifier.

4.2 RELATIONSHIP BETWEEN THE CC AND NNC

It is a simple matter to show that the CC is a homopolar version of the modified NNC [1]. This can be achieved by comparing equations (3.11) and (3.19). Both of these classifiers produce piece-wise hyper-spherical decision surfaces. However the NNC is more flexible since it can realise any such surface, whereas the CC can only achieve a subset of these (compare Figures 3.12 and 3.15). The CC is capable of recognising members of a relatively dense, compact class which is embedded in another, more diffuse class. It would be more natural (requiring fewer locates) to use the modified NNC when the classes are both diffuse (Figure 4.1). Since the modified NNC can realise any surface which can be produced by a CC, it would seem natural to discard the latter here. However, as we shall see in Chapter 6, the CC is very easily implemented in electronic hardware. Moreover, the modified NNC suffers from the disadvantage that under certain conditions its decision surface is critically dependent both upon the positions of its locates,

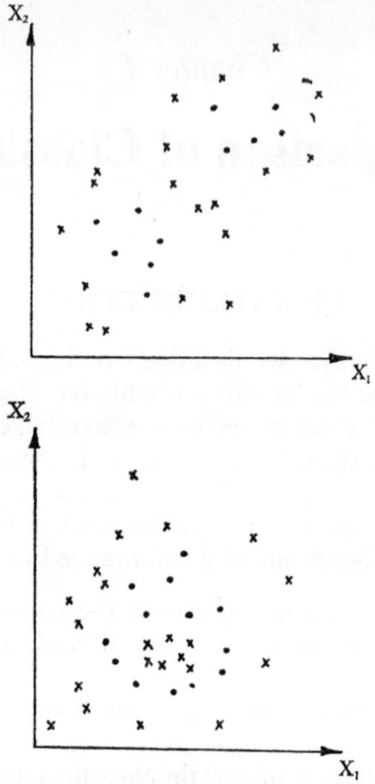

Figure 4.1 Showing pattern distributions suiting different classifiers. (A) a CC or modified NNC can separate these classes easily. (B) these classes can easily be separated by a modified NNC but not by a CC.

and the values of its weights. (We shall return to this topic later in this chapter.)

4.3 RELATIONSHIP BETWEEN THE NNC AND THE HETEROPOLAR PFC

Our major objective here is to provide an analytic basis for the assertion that the NNC and PFC are simply variations of the same general classifier [1]. We shall show that equation (3.13), which defines the general PFC, allows us to approximate the surface of the modified NNC to whatever accuracy we may choose.

First, we shall remind ourselves that the equation of the general PFC is given by equation (3.13):

$$M = \operatorname{sgn}\left\{T + \sum_{j=1}^{N_{+1}} P(D_{+1,j})W_{+1,j} - \sum_{j=1}^{N_{-1}} P(D_{-1,j})W_{-1,j}\right\} \tag{3.13}$$

We shall assume that the elements of the **X**-vector are all quantised, and hence that the $D_{i,j}$ are also quantised. Let the smallest quantisation increment on $D_{i,j}$ be d. Now we shall choose the "potential function", $P(D)$, to be a very rapidly changing function, such that:

$$P(D + d)W_{\max}\max(N_{-1}, N_{+1}) \ll P(D)W_{\min} \tag{4.1}$$

where $W_{\min}$ and $W_{\max}$ are the extreme values of the weights. Since $P(D)$ is changing so rapidly we can *usually* replace the sum,

$$\sum_{j=1}^{N_i} P(D_{i,j})W_{i,j} \tag{4.2}$$

by its largest term

$$\max_{j=1,\ldots,N_i} (P(D_{i,j})W_{i,j}) \tag{4.3}$$

When two or more of the $D_{i,j}$ are equal we *may* not be able to make this approximation.

The only points at which this approximation would seem to be invalid lie on the decision surface of the NNC, where its decisions are not defined. By substituting (4.3) for (4.2) in equation (3.13) we obtain the following (we are also setting T to zero):

$$M = \operatorname{sgn}\left\{\max_{j=1,\ldots,N_{+1}} (P(D_{+1,j})W_{+1,j}) - \max_{j=1,\ldots,N_{-1}} (P(D_{-1,j})W_{-1,j})\right\} \tag{4.4}$$

Now we shall add one further constraint on $P(D)$. We shall require that for all positive values of a:

$$P(Da) = P(D)P(a) \tag{4.5}$$

Furthermore, we shall define $P^*(D)$ as the inverse function of $P(D)$ so that;

$$P^*(P(D)) = D \tag{4.6}$$

Since $P(D)$ is a monotonically decreasing function, maximising $P(D_{i,j})W_{i,j}$ is equivalent to minimising $P^*(W_{i,j})D_{i,j}$. Equation (4.4) will therefore give the same values for M as will the following equation;

$$M = -\operatorname{sgn}\left\{\min_{j=1,\ldots,N_{+1}} (D_{+1,j}P^*(W_{+1,j})) - \min_{j=1,\ldots,N_{-1}} (D_{-1,j}P^*(W_{-1,j}))\right\} \tag{4.7}$$

Equation (4.7) has the same form as (3.11), which defined the modified NNC. We have shown then that the modified NNC produces decisions indentical with those from a PFC with a rapidly decaying potential function, $P(D)$. The only significant constraint we have added is that $P(D)$ obeys (4.5).

The quantisation of the elements of the **X**-vector is not important, since we can tolerate fine resolutions by making $P(D)$ change more rapidly.

We have seen that the modified NNC and the heteropolar PFC are closely related, and that under certain conditions their decision surfaces are identical. Even if $P(D)$ changes more slowly than required by inequality (4.1), the decision surfaces of the PFC and modified NNC are similar [2]. Proving this analytically is far less satisfactory than simply plotting surfaces, which we do in Figures 4.2 to 4.4.

We have obtained similar results in a large number of other experiments.

We have shown above that the modified NNC has a similar decision surface to the heteropolar PFC *which has T equal to zero*. We can produce a further modification to the NNC which yields a similar decision

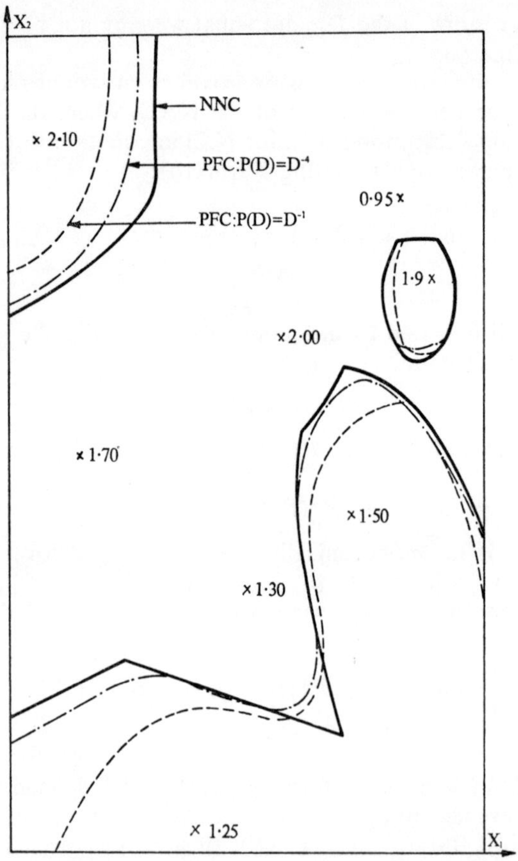

Figure 4.2 Comparison of decision surfaces produced by the modified NNC and two PFCs. All three classifiers used the same locates (indicated by circles and crosses) and weights (numbers). This diagram was traced from a computer print-out, hence the vertical axis is elongated by a factor of 5:3, compared to the horizontal. Reproduced by permission of American Elsevier Publishing Company Inc.

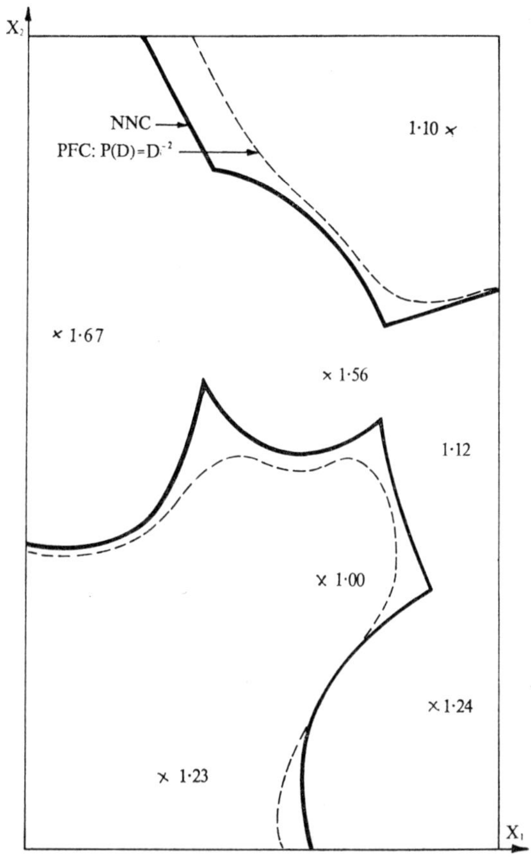

Figure 4.3 Comparison of NNC and PFC decision surfaces. (For details, see Figure 4.2.) Reproduced by permission of American Elsevier Publishing Company Inc.

surface to the PFC with finite T. To do this we must add a constant term into the equation of the modified NNC.* This does not seriously alter the decision surface which is still piece-wise hyperspherical.

4.4 RELATIONSHIP BETWEEN THE CC AND HOMOPOLAR PFC

By now the reader will probably not be surprised to hear that there is a close relationship between the CC and the homopolar PFC. We have

* The equation is then,

$$M = \operatorname{sgn}\left\{T' + \min_{j=1,\ldots,N_{-1}} (W_{-1'j}D_{-1,j}) - \min_{j=1,\ldots,N_{+1}} (W_{+1,j}D_{+1,j})\right\}$$

where T' is the constant term.

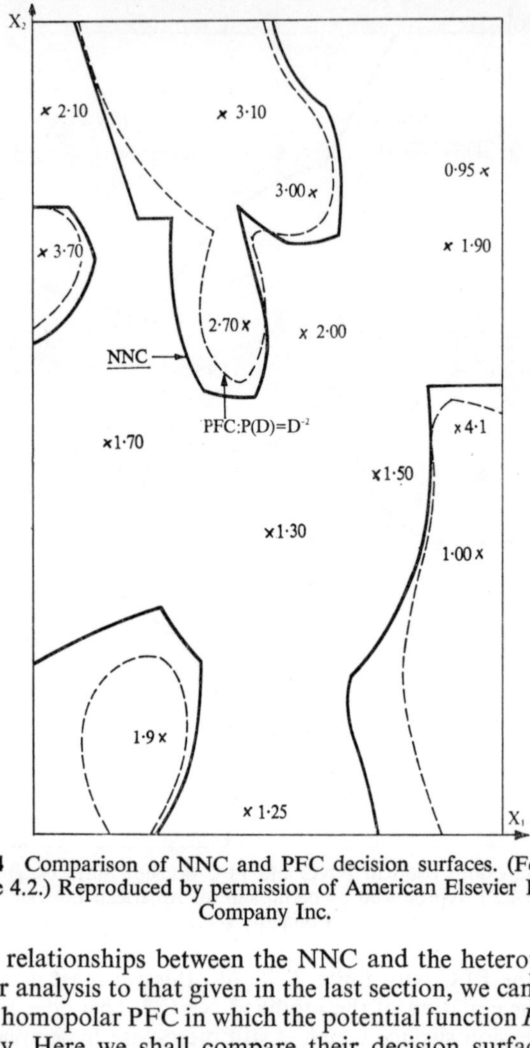

Figure 4.4 Comparison of NNC and PFC decision surfaces. (For details, see Figure 4.2.) Reproduced by permission of American Elsevier Publishing Company Inc.

established relationships between the NNC and the heteropolar PFC. By a similar analysis to that given in the last section, we can prove that the CC is a homopolar PFC in which the potential function $P(D)$ decays very rapidly. Here we shall compare their decision surfaces visually (Figures 4.5 and 4.6). Notice the tendency of the homopolar PFC to create "necks" between the islands. The decision surface of the homopolar PFC is smoother, and generally more rounded than that produced by the CC.

4.5 SUMMARY SO FAR

Our conclusions about the relationships between classifiers are distributed throughout this and the previous chapter. It seems appropriate to collect these together now, and try to form a coherent picture of our

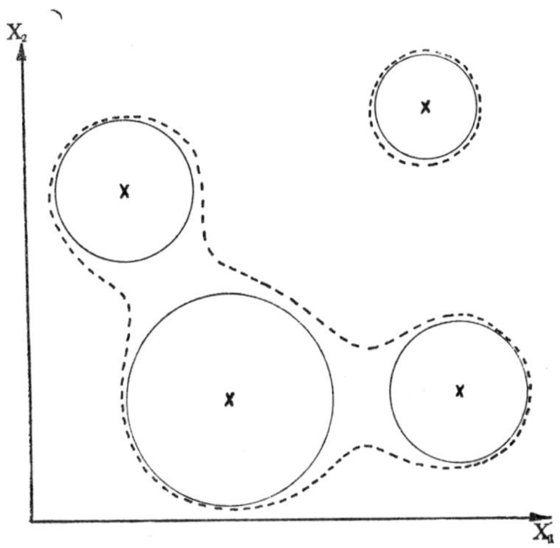

Figure 4.5 Comparison of the decision surfaces of a CC and homopolar PFC. The same locates (indicated by crosses) were used.

subject. We have incorporated our major points into a diagram (Figure 4.7). Letters in this figure indicate the nature of the relationship, the key is listed below:

(A) hetero-/homopolar pairs.
(B) the NNC/CC are special cases of the hetero-/homopolar PFC. The decision surfaces of the NNC and heteropolar PFC are similar, and those of the CC and homopolar PFC are similar.

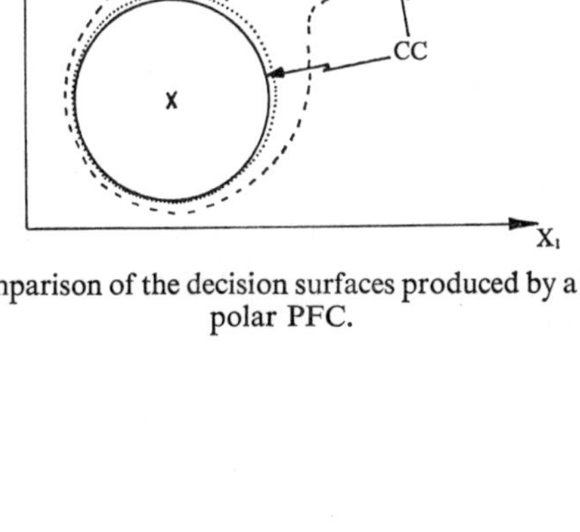

Figure 4.6 Comparison of the decision surfaces produced by a CC and homopolar PFC.

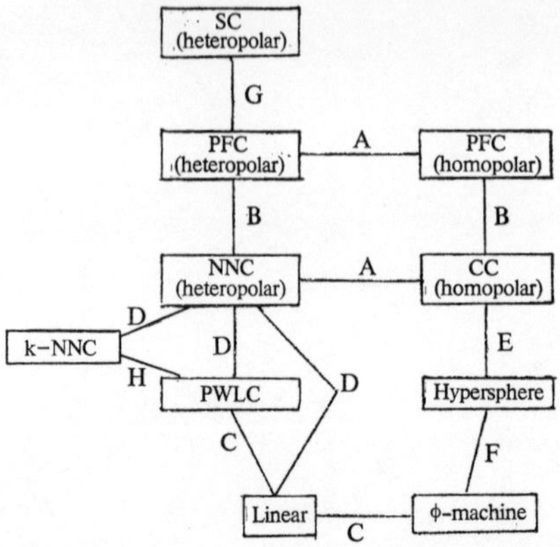

Figure 4.7 Illustrating the relationship between the classifiers. The key is included in the text.

(C) The linear classifier is a major component of both the PWLC and ϕ-machine. Its decision surface can be realised by both.
(D) The NNC can realise any linear or piece-wise linear decision surface, but can realise other surfaces which are not piece-wise linear. The NNC can also realise any k-NNC surface, which is piece-wise linear.
(E) The hypersphere classifier is the major component of the CC.
(F) The ϕ-machine can realise the hypersphere classifier.
(G) The SC is a special case of the heteropolar PFC. Their decision surfaces are similar.
(H) The k-NNC surface can be realised by a PWLC.

4.6 INSTABILITY

Under certain conditions the performance of a classifier may be critically dependent upon the values of its stored parameters. It may be possible to reclassify more than half of the descriptor space by the infinitesimal variation of only one parameter. This effect, called instability, has been found in many classifiers. It is not possible to define the term exactly. Some classifiers, for example the NNC, exhibit an instantaneous change of decision surface. In other classifiers the decision surface changes very rapidly, but continuously. Both of these effects will be described as instability. The difference is unimportant in practice.

Instability is important for two reasons:

(a) Learning is more difficult; there may be no smooth approach to the optimal solution.
(b) Electrical noise, quantisation noise or component variations could cause sudden, violent changes in the performance of the classifier.

There are several types of instability resulting from variations of the locates, weights and threshold. Some instability can be avoided by using the Euclidean distance. For our present purposes we include the Sebestyen Classifier as a PFC.

4.7 INSTABILITY DUE TO LOCATES MOVING

The NNC (and hence the heteropolar PFC as well) suffers from two types of instability resulting from variations of the locates.

4.7.1 The First Type of Instability

The first type occurs whichever distance is used. If the locates are close together, a small perturbation of one of them may result in a large variation in the position of the decision surface (Figure 4.8). This

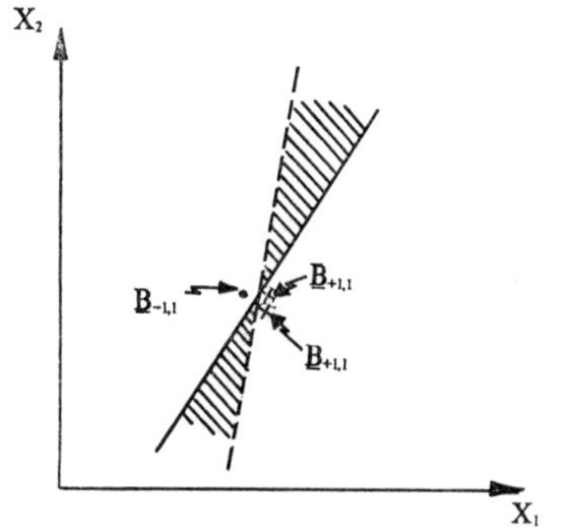

Figure 4.8 Showing the effects of noise in a NNC which has two close locates. By a very small change in the positions of the locates, the decision surface can be moved through an appreciable angle. When the locates are $\mathbf{B}_{-1,1}$ and $\mathbf{B}_{+1,1}$ the decision surface is as shown by the solid line. When the locates are at $\mathbf{B}_{-1,1}$ and $\mathbf{B}'_{+1,1}$ the decision surface moves to the position indicated by the broken line. This is type 1 instability.

type of instability would not be important if we could develop training procedures which would always keep the locates far apart.

4.7.2 The Second Type of Instability

This type of instability is particularly associated with the use of non-Euclidean distances [3]. When the locates have certain relative orientations the decision surface can change suddenly. (Figures 4.9 to 4.11).

4.8 INSTABILITY DUE TO WEIGHTS VARYING

4.8.1 The Third Type of Instability

This type of instability occurs in the modified NNC and heteropolar PFC when the weights have nearly the same value. (Figure 4.12). This effect occurs whichever distance is used.

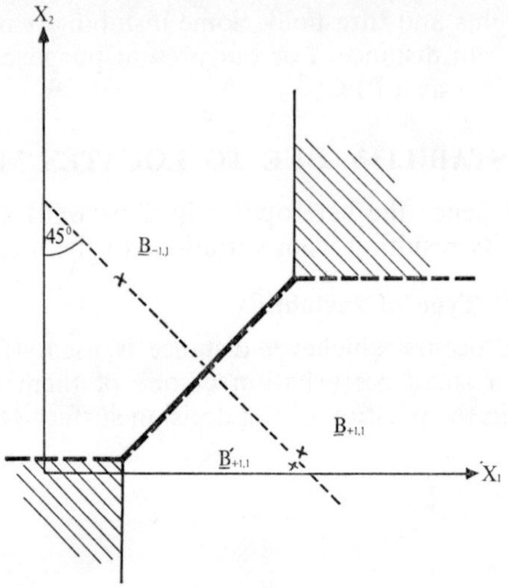

Figure 4.9 Showing the effect of type 2 instability in a NNC which uses the city block distance. When locates $\mathbf{B}_{-1,1}$ and $\mathbf{B}_{+1,1}$ are used, the decision surface is given by the solid curve. When the locates are at $\mathbf{B}_{-1,1}$ and $\mathbf{B}'_{+1,1}$ the decision surface is as shown in the broken curve. The cross-hatched region indicates that part of the space that is reclassified. The effect of locate variation is far less serious if the locates are well separated. (source [3].)

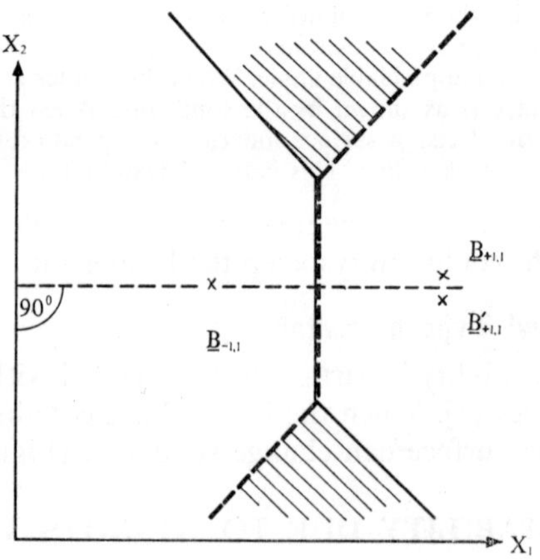

Figure 4.10 Type 2 instability in a NNC using the square distance. Other details are as in Figure 4.9. (source [3].)

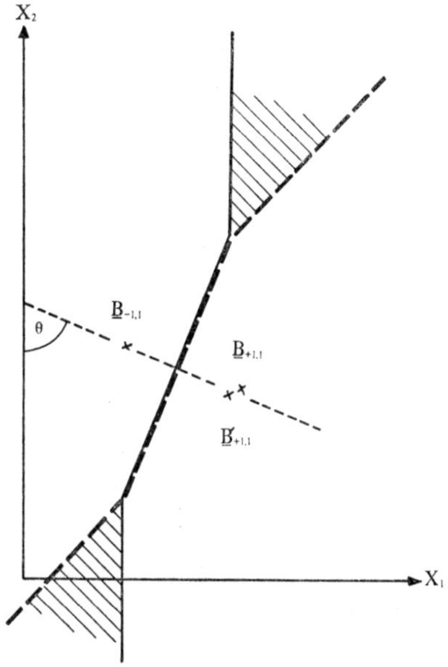

Figure 4.11 Type 2 instability in a NNC using the octagonal distance. Instability occurs when θ is $22\frac{1}{2}°$ or $67\frac{1}{2}°$. Other details are as in Figure 4.9. (Source [3].)

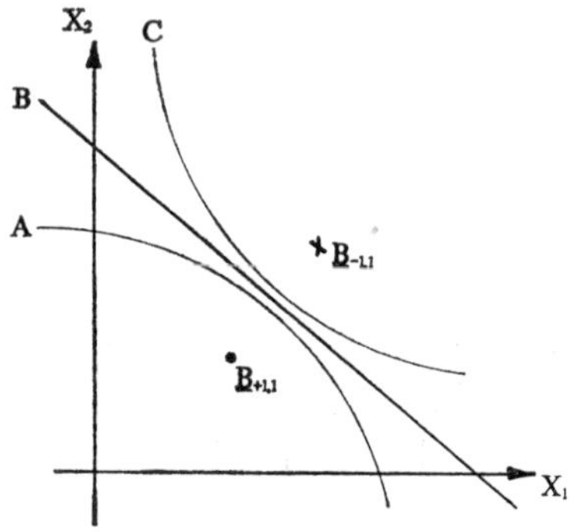

Figure 4.12 Type 3 instability in a modified NNC using the Euclidean distance. When the weights are equal the decision surface is as shown in (B). When $W_{+1,1}$ is greater than $W_{-1,1}$ the decision surface is at (C). When $W_{+1,1}$ is the smaller, the surface is at (A).

4.9 INSTABILITY DUE TO THRESHOLD VARIATIONS

Two further types of instability occur in the PFC but have no counterpart in the NNC or CC. Both arise because the "potential" is nearly constant over a significant part of the descriptor space.

4.9.1 The Fourth Type of Instability

This type of instability is found in both the homo- and heteropolar PFCs. To describe it we must consider the potential (S):

$$S = \sum_{j=1}^{N_{+1}} P(D_{+1,j})W_{+1,j} - \sum_{j=1}^{N_{-1}} P(D_{-1,j})W_{-1,j} \tag{4.8}$$

At points which are far removed from all of the locates, each of the $D_{i,j}$ is large, and hence all of the $P(D_{i,j})$ are small. Hence S is also small and the decision, M, which is given by

$$M = \text{sgn}\ (S + T) \tag{4.9}$$

is sensitive to changes in T. (Figures 4.13 to 4.15). The fourth type of instability occurs whichever distance is used.

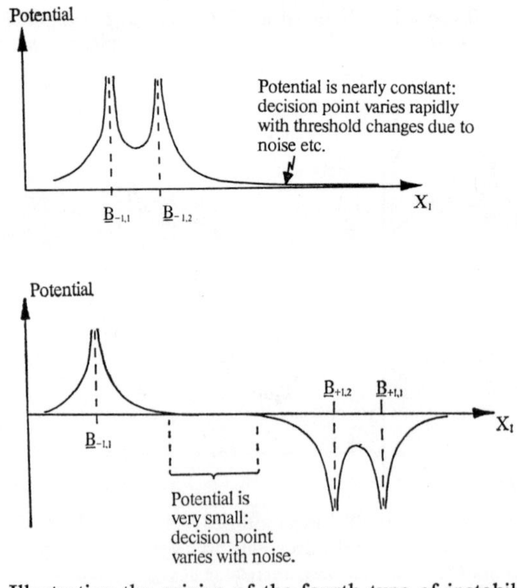

Figure 4.13 Illustrating the origins of the fourth type of instability in (A) a homopolar PFC, and (B) a heteropolar PFC. For convenience of display a single dimensional descriptor space is shown. A similar situation holds in a higher dimensional space.

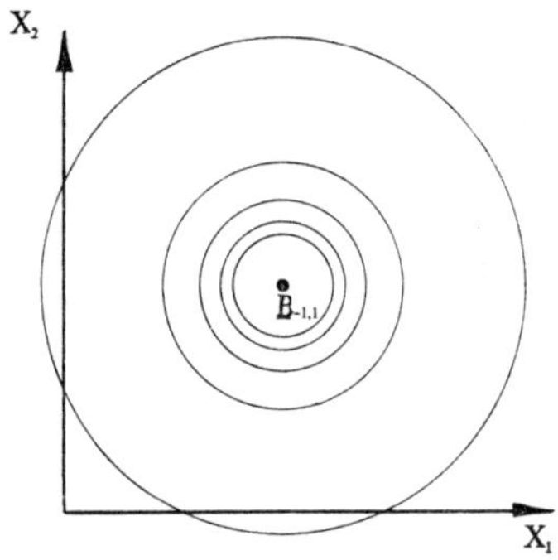

Figure 4.14 Type 4 instability in a homopolar PFC, which has a single locate, $\mathbf{B}_{-1,1}$. The circles represent equally spaced contours of potential. The next contour would be at infinity.

4.9.2 The Fifth Type of Instability

This form of instability is associated with the use of non-Euclidean distances in homopolar PFCs. The potential hills have plateaux with flat tops (Figures 4.16 and 4.17).

4.10 SIXTH TYPE OF INSTABILITY

Type 6 instability affects the ϕ-machine. To illustrate this, consider a ϕ-machine which calculates a decision from the expression $\text{sgn}(ax_1^2 + bx_2^2 + c)$. If $ab > 0$ the decision surface is an ellipse in (x_1, x_2)-space, while it is a hyperbola if $ab < 0$. In this case, instability occurs when $ab = 0$.

4.11 SUMMARY OF THE CHAPTER

The most important point arising from sections 4.2–4.5 is that there is little difference between the capabilities of the modified NNC and

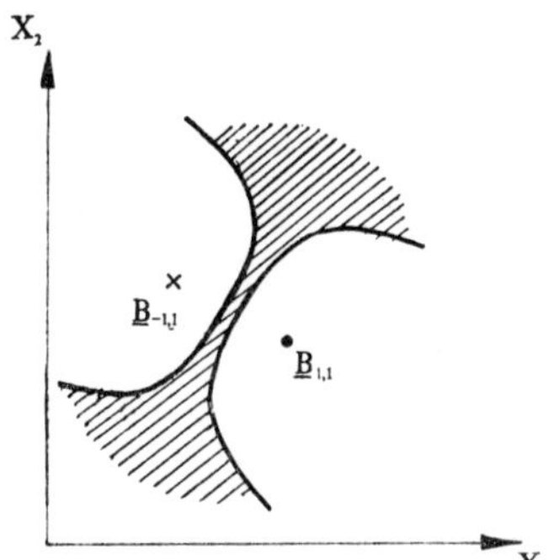

Figure 4.15 Type 4 instability in a heteropolar PFC. The two solid curves represent the extreme positions of the decision surface during instability.

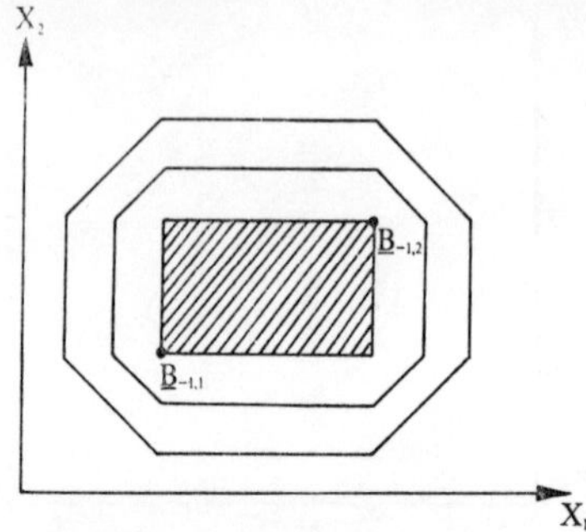

Figure 4.16 Type 5 instability in a homopolar PFC which uses the city block distance, and has two locates, $\mathbf{B}_{-1,1}$ and $\mathbf{B}_{-1,2}$.

heteropolar PFC, nor between the CC and homopolar PFC. (The SC is a special case of the heteropolar PFC and will therefore be included in this, and any other general remarks we make about the PFC.) The PWLC can be regarded as being equivalent to the (unmodified) NNC, since these classifiers can always match each others decision surfaces exactly. Hence we can eliminate the PWLC from further discussion, since it is simply an alternative method of realising the NNC. However the PWLC is not capable of matching the modified NNC. The linear classifier is also a special case of the NNC. The ϕ-machine has no close parallels with the PFC/NNC family.

Instability has provided some important clues as to which classifiers we should eliminate (Table 4.1). The classifiers do not suffer linear and compound from instability of any sort, while the ϕ-machine is only affected by one type (6).

The modified NNC suffers from 3 types of instability (1, 2, 3). The heteropolar PFC is worse with 4 types (1, 2, 3, 4). The homopolar PFC

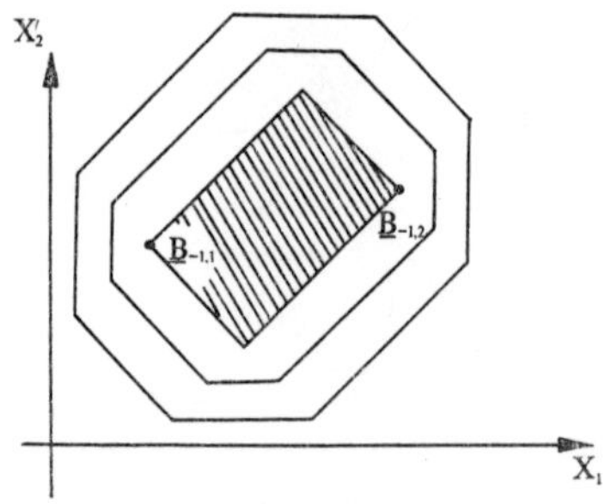

Figure 4.17 Type 5 instability in a homopolar PFC which uses the square distance, and has two locates, $\mathbf{B}_{-1,1}$ and $\mathbf{B}_{-1,2}$.

Table 4.1 Summary of causes of instability in classifiers

Instability	Classifier	Distance	Comments
1	NNC Hetero. PFC	Any	Avoided if locates are far apart.
2	NNC Hetero. PFC	Non-Euclidean	Pronounced with city block and square distances.
3	Modified NNC Hetero. PFC	Any	—
4	Homo. PFC Hetero. PFC	Any	Avoided if locates fill space.
5	Homo. PFC	Non-Euclidean	—
6	ϕ-machine	—	—

suffers from instability of types 4 and 5. It is important to notice that type 2 instability can be completely avoided by using the Euclidean distance. Type 1 instability in a NNC or PFC can be avoided if we keep the locates well separated during learning. However, this tends to make type 4 instability worse in a PFC. We can therefore narrow the choice to the following:

compound classifier (any distance),
modified NNC (using Euclidean distance),
ϕ-machine.

Let us briefly review the relative merits of these classifiers again.

(a) The CC does not suffer from instability. On the other hand it does not produce such flexible decision surfaces as the modified NNC for a given store size.
(b) The modified NNC suffers from type 3 instability, which is unavoidable. It is also subject to type 1 instability which may be avoided, if we can develop suitable training rules. However, this might lead to a more expensive training programme.
(c) The ϕ-machine does not make full use of the natural clustering of patterns in the descriptor space. Hence, it would be difficult to achieve a separation between a class containing say, 3 compact clusters, embedded in a mass of patterns from another class. Such a task is, of course, natural for a CC or NNC. Another important feature of the ϕ-machine is the problem of finding the appropriate polynomial terms. It is impossible to perform an exhaustive search, and so far no really effective solution has been described [4, 5].

We have reduced our choice to the ϕ-machine, CC, and modified NNC. In the following chapter, we concentrate mainly on the CC, as this classifier has the merit of having the best developed training rules.

REFERENCES

1. B. G. Batchelor, N. L. Ford and B. R. Wilkins, Family of Pattern Classifiers, *Electronics Letters*, **6**, 12, 386–389, 1970.
2. B. G. Batchelor, Comparison of the Decision Surfaces of the Nearest Neighbour and Potential Function Classifiers, *Information Sciences*, **5**, 171–178, 1973.
3. B. G. Batchelor, Instability of the Decision Surfaces of the Nearest Neighbour and Potential Function Classifiers, *Information Sciences*, **5**, 179–186, 1973.
4. A. N. Mucciardi and E. E. Gose, Evolutionary Pattern Recognition in Incomplete Non-linear Multi-threshold Networks, *Trans. I.E.E.E., EC-15*, 2, 257–262, 1966.
5. B. R. Wilkins and B. G. Batchelor, Evolution of a Descriptor Set for Pattern Recognition, *Proc. Conference on Technical and Biological Problems of Control*, Instrument Society of America, 794–799, 1970.

Chapter 5

Learning

5.1 INTRODUCTION

So far we have largely ignored the problem of finding suitable values for the stored parameters required by our classifiers. This will form our next subject for discussion. It is impossible, in a short chapter such as this, to discuss more than a few of the many hundreds of learning schemes that have been proposed.* We have been able to show in Chapter 4 that the ϕ-machine, modified NNC and CC, are likely to be the most useful classifiers. It seems reasonable, therefore, to concentrate here on the development of learning rules for these classifiers. However, we cannot even do full justice to all of the available learning schemes for these three classifiers. The author has therefore selected what he believes to be the most useful learning rules.

Suppose that we are designing a CC. There are then three questions that we must answer:

(a) How many locates do we require?
(b) Where should we place the locates in the descriptor space?
(c) What values should we use for the size parameters?

Similar questions arise when we attempt to design a modified NNC (or a PFC). When our classifier is a ϕ-machine we are faced with the following problems:

(d) How many polynomials terms do we require?
(e) What should the polynomial terms be? (For example, should the discriminant function contain a term in $X_1X_2^2$?)
(f) What values should we use for the weights?

Most of our learning rules are based on intuition and have not yet been proved to converge using mathematical analysis. Whereas we should welcome such a proof, we do not feel that its absence should deter us from pressing on with our experimental study of intuitively-based schemes. Our most frequent method of demonstrating convergence to the reader will be to ask him to compare the decision surface, obtained by learning, with either the teacher decision surface, or a scattergram display of the training set. In a high-dimensional space

* Other books cover a wider range, usually at the expense of omitting practical detail, which we feel is very important [1–5].

this approach is impossible. We must then resort to quoting initial and final error rates to show that our rules are effective.

5.2 LEARNING AS A FORM OF FEEDBACK

Most of our training procedures have an iterative structure (Figure 5.1) in which the classifier is slowly modified in the light of experience gained about its deficiencies. This is a typical feedback arrangement, and has a number of important advantages.

(a) The effect of any one pattern is small. This is important since patterns which are unrepresentative of their class would otherwise cause gross changes in the performance of the classifier.
(b) Pattern vectors are presented serially to the classifier. Hence, only one pattern vector* need be stored at any one time. A large training set may contain up to 10^6 patterns.
(c) We know, prior to finding parameters for the classifier, what functions it will compute. This is important since these functions have been specially selected because they are convenient for computation.
(d) Errors in a hardware classifier can be automatically compensated by incorporating it into a feedback loop. (Such errors may be due to component selection tolerances.)
(e) Feedback allows considerable flexibility in the parameter adjustment rules, which may therefore be chosen by considering their realisation in hardware/software.

5.3 METHODS OF ADAPTATION

We shall find it helpful to introduce a number of terms to describe various types of learning. These are listed below.

(*a*) *Variable storage learning*

These are methods in which the quantity of information stored by a classifier is variable. Only store generation and deletion are allowed; once set, the contents of a store cannot be modified. Thus, in a NNC we could add and remove locates, but could not move them around in the descriptor space.

(*b*) *Parameter adjustment techniques*

These are methods in which the store size is fixed and store contents are variable. In a CC, we could move the circles around and adjust their radii, but the total number of circles would remain unaltered.

(*c*) *Growing*

This is a procedure combining parameter adjustment and variable storage techniques. The classifier size increases.

* This does not include the locates which are permanently stored within the classifier.

(*d*) *Pruning*
This technique is similar to growing, except that the classifier size is reduced.

(*e*) *Error-correction*
This is a method of parameter adjustment which is executed following an error by the classifier (Figure 5.1B).

(*f*) *Reinforcement*
This is a method of adaptation which is executed following a correct decision by the classifier (Figure 5.1B).

5.4 STORE ALL, A METHOD OF DESIGNING A NNC (OR PFC) BY STORING THE COMPLETE TRAINING SET

We begin our development of training rules by considering perhaps the simplest method of finding locates in a NNC, PFC or k-NNC. In the store-all rule, every member of the training set is used as a locate in one of these classifiers. The major objection to this procedure is that a classifier designed in this way has a lot of redundant locates (Figure 5.2). As a result, the procedure is quite impractical for all but the simplest problems.

5.4.1 Estimating the Bayes' Error Probability

At this point, we shall digress briefly to consider the use of the store-all rule to assess the quality of the training set. The error rate of the NNC designed by a store-all rule can be related to the error rate of the Bayes' classifier. The latter classifier can only be realised if we know the probability density functions for both pattern classes. However, its error rate (β) can be estimated experimentally. The following inequality was established by Cover and Hart [6], and relates β to the error rate (η) of an infinitely large NNC that has been designed by the store-all rule. β and η are related as follows,

$$\tfrac{1}{2} - \tfrac{1}{2}\sqrt{1 - 2\eta} < \beta < \eta \tag{5.1}$$

Hence, if we can find η, we can calculate bounds for β. In practice, we can only find an approximate value for η, since our NNC must be finite. In order to ensure that we obtain an accurate estimate for η we suggest that the NNC be enlarged until our estimate is nearly constant. In Figure 5.3 we outline a practical procedure for estimating β which makes use of this technique for obtaining accurate results.

We now have an absolute standard against which we can compare the results obtained from a practical classifier. This is an important facility for two reasons:

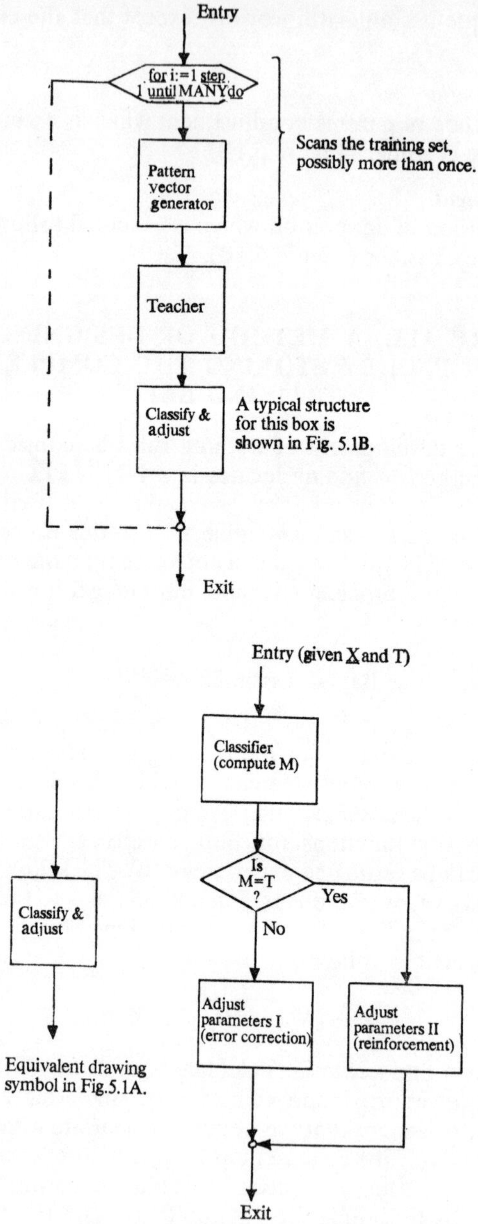

Figure 5.1 Flowchart of an iterative learning scheme employing feedback. (A) showing iterative structure. (B) showing error-correction and reinforcement learning.

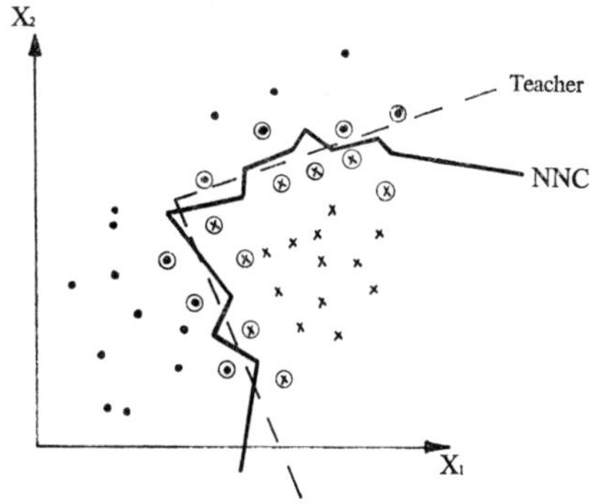

Figure 5.2 Performance of the store-all rule for designing a NNC. The points were classified by an artificial teacher, using the decision surface shown by the broken curve. In the store-all rule, every member of the training set becomes a locate in the NNC. However, some of the locates generated in this way are redundant. The locates contributing to the NNC decision surface (solid curve) are encircled. 40 locates are shown here, but only 15 are useful, while only 3 are needed to specify the teacher decision surface.

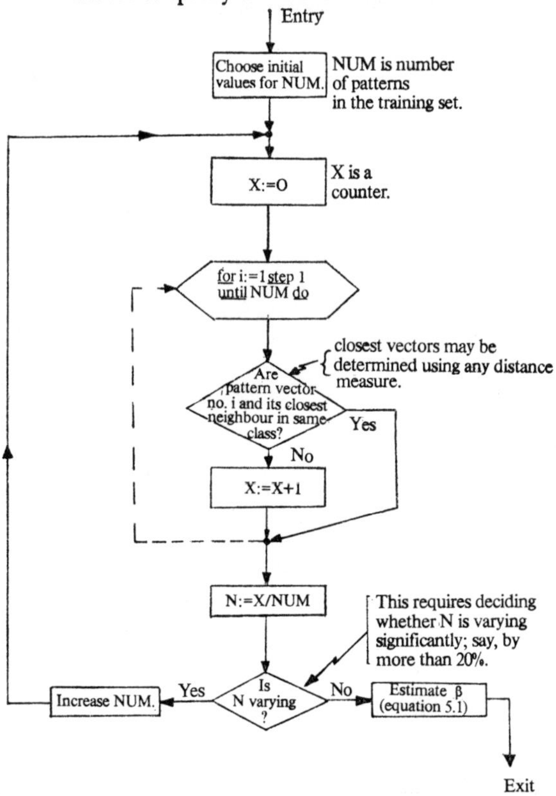

Figure 5.3 Procedure for estimating β, the error rate of the Bayes' classifier.

(a) Since our classifier design procedures have not yet been made respectable by mathematical analysis, we shall find it an advantage to be able to compare the practical classifier with the best attainable classifier (Bayes'). This is particularly valuable in convincing customers that our classifier is nearly optimal.

(b) During our classifier design program, we often need to know whether it is likely to be worthwhile increasing the number of locates, or whether it would better to start eliminating redundant locates. The ability to estimate β would be of great value in this decision.

5.5 ERROR CORRECTION LEARNING IN THE LINEAR CLASSIFIER AND ϕ-MACHINE

5.5.1 Learning in a Linear Classifier

Despite its relative inflexibility compared to, say, the NNC, the linear classifier is still widely discussed in the literature. As a result, there have been many suggestions for methods of finding suitable

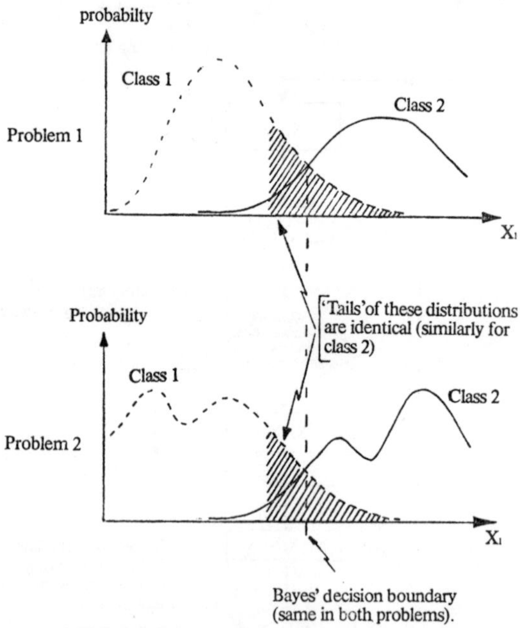

Figure 5.4 Shows probability distributions for two problems which have the same Bayes' decision boundary. Reinforcement learning would produce different solutions, since the decision boundary is sensitive to the typical patterns, close to the peaks of the probability "hills". On the other hand, error-correction learning largely ignores these and places greater emphasis on the "tails" (shown as shaded for class 1).

values for the weights. (In 1968 the author found eleven without difficulty.) We have selected just two for discussion here. Most, but not all, of these suggestions use error-correction learning. The reason is that reinforcement learning places greater emphasis on patterns which are typical of their class, compared to error correction learning, which largely ignores these. Error-correction methods operate on patterns close to the optimal decision boundary (Figure 5.4).

Figure 5.5A defines an error-correction training rule for a linear classifier. The quantity k is called an *iteration parameter*. Its value is relatively unimportant provided that it is positive, and does not increase with time. It may be constant or reducing with time. A value of unity is, of course, particularly convenient. Figure 5.6 shows how the decision surface is moved in response to a single error. If a solution exists,* this rule will always achieve convergence in a finite time [2].

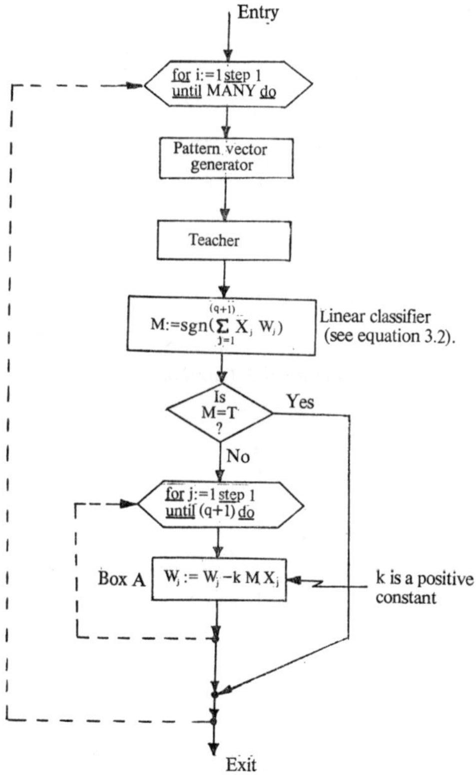

Figure 5.5 (A) Procedure for designing a linear classifier by error-correction learning. k is a positive constant, called an iteration parameter. (B) Procedure for designing a ϕ-machine, by error-correction learning. Notice the strong similarity between this and (A).

* This is one in which the error rate is zero.

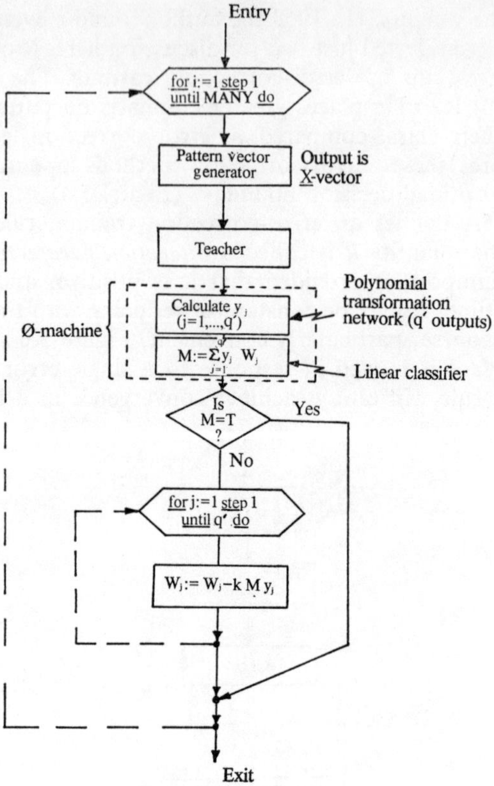

Figure 5.5 (*continued*)

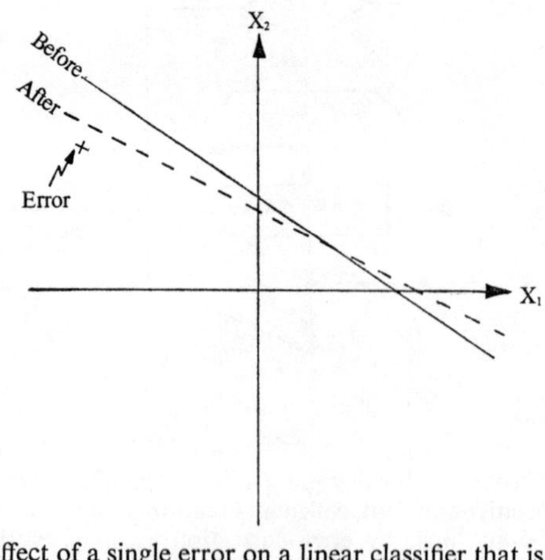

Figure 5.6 Effect of a single error on a linear classifier that is being designed by the procedure of Figure 5.5. The modification is greatly exaggerated for clarity of display.

Experimental tests have shown that if there is no perfect solution, this rule will still place the linear decision surface close to the optimal position (Figures 5.7 and 5.8), [7].

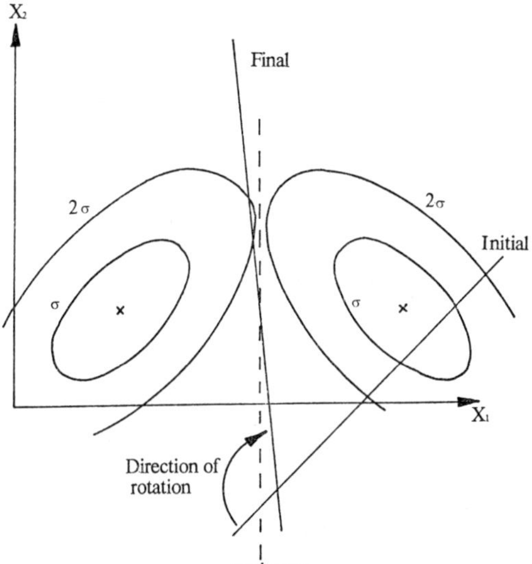

Figure 5.7 Result of designing a linear classifier by the procedure of Figure 5.5. The ellipses indicate contours of the probability density functions. Both classes contained a single Gaussian cluster. Type B experiment.

Our second variation of the training rules is included to emphasise that it is possible to change a feedback procedure without seriously disturbing its performance. Consider Box A in Figure 5.5A, which required that we replace the weight W_i by

$$W_i - kMX_i \tag{5.2}$$

To avoid the analogue multiplication inherent in (5.2), W. K. Taylor [8] suggested the following alternative rule for modifying W_i:

$$W_i := W_i - kM\text{sgn}(X_i) \tag{5.3}$$

The modifications to the W_i then all have the form

$$W_i = W_i \pm k \tag{5.4}$$

since both M and $\text{sgn}(X_i)$ may only assume the values ± 1. The simpler rule (for implementation) defined by (5.3) might be used to replace Box A in Figure 5.5. The surprising result is that there is very little change in the performance of the learning rule (Figure 5.9).

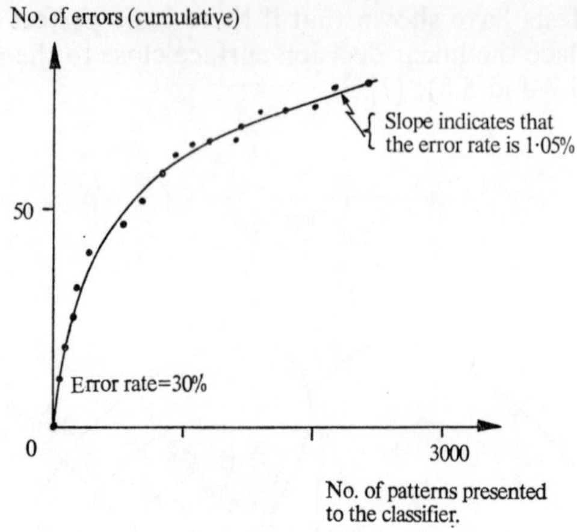

Figure 5.8 Designing a linear classifier to distinguish English and Latin prose, using letter frequencies as descriptors. Further detail is provided in section 7.7, and [7, 9].

5.5.2 Learning in a φ-Machine

The ϕ-machine may be taught in just the same way as the linear classifier, with one minor alteration (Figure 5.5B). This interchange of learning rules is possible because both classifiers compute a linear function of their weights. Let us briefly consider a ϕ-machine consisting

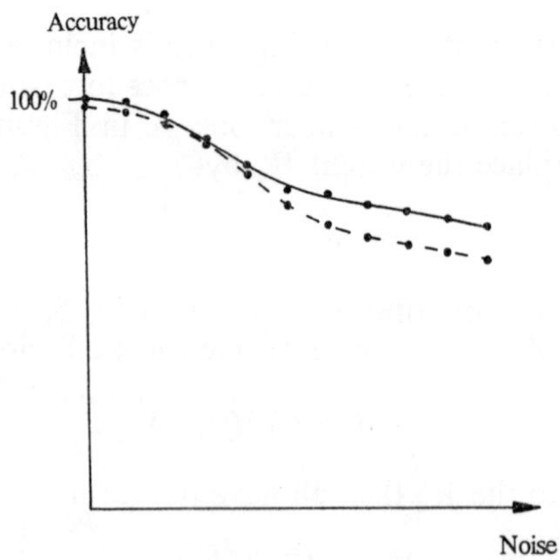

Figure 5.9 Effect of noise on learning in a linear classifier. This was a type C experiment (section 2.8.3). "Accuracy" is measured by the cosine of the angle between the weight vectors of the optimum and learning classifier (after 5000 patterns have been examined). The solid curve relates to the results obtained using the simplified modification rule of Equation 5.3. The broken curve relates to the procedure defined in Figure 5.5. (Further details are given in [7].)

of a polynomial transformation network, P, followed by a linear classifier, L. Let the analyser be A. Our pattern recognition system consists then of an analyser, A, followed by a ϕ-machine, $(P + L)$. Now we can associate P with A to produce a new, extended analyser, $(A + P)$. The system then contains an analyser $(A + P)$, and a classifier L. L can of course be trained by the rules defined in Figure 5.5A. However, notice that the inputs to L are the outputs of P (Y_i in section 3.3) and not the outputs from A (X_i in section 3.3). Hence, the weight modification rule (Box A in Figure 5.5A) must be replaced by the following (see Figure 5.5B):

$$W_i := W_i - kMY_i \tag{5.5}$$

We have so far ignored the difficult problem of selecting suitable terms for the polynomial transformation network. Sebestyen [1] considers another technique for designing a ϕ-machine. In this procedure the complexity of the polynomial transformation network is gradually increased. Unfortunately Sebestyen's procedure requires very large amounts of computation when there are many dimensions and/or high degree polynomial terms are considered. Another technique was studied by the present author [9]. This procedure was developed from a similar idea by Mucciardi and Gose [10]. In this procedure "useless" inputs (those with a very small weight) were discarded from a linear classifier, while new inputs were selected at random. The technique worked well on a problem in which the descriptors were nearly independent, but failed to do so when there was a high level of correlation between them. This situation exists, of course, in a ϕ-machine, which might, for example, compute $Y_1 = X_1X_2$ and $Y_2 = X_1X_2^3$. We are left then with the problem of designing a ϕ-machine being effectively unresolved.

5.6 ERROR CORRECTION LEARNING IN A HYPERSPHERE CLASSIFIER

In the following section we shall describe learning rules for a CC. To prepare us for this, we shall find it convenient first to discuss methods for training a hypersphere classifier. This is of course, the simplest possible CC, having just one subclassifier.

Error-correction learning rules for a hypersphere classifier are defined in Figure 5.10. The iteration parameters, k and l, would typically have the values:

$$0.0001 \leq k \leq 0.05$$

$$0.001\lambda \leq l \leq 0.05\lambda$$

where λ is the length of the diagonal of the smallest window containing the pattern points in the descriptor space (Figure 5.11). The values of k and l are not critical. Smaller values may be used, although this will increase the learning time. On the other hand, larger values should be

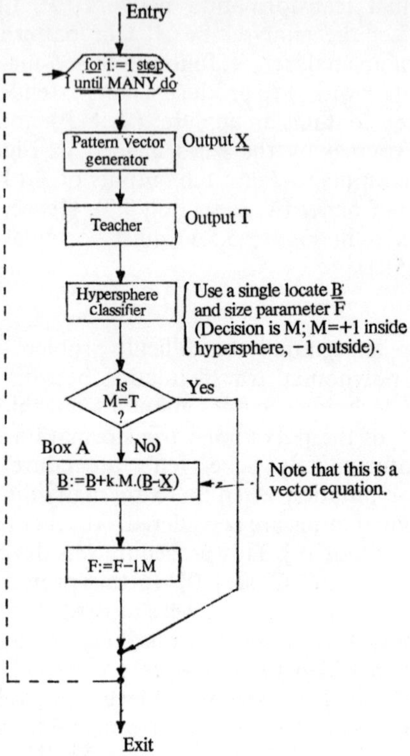

Figure 5.10 Procedure for designing a hypersphere classifier by error-correction learning. Notice the simplified notation. **B** is the locate, and F the size parameter. k and l are iteration parameters.

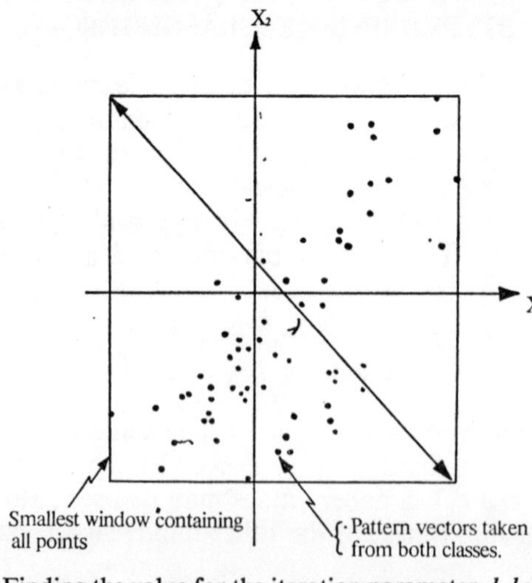

Figure 5.11 Finding the value for the iteration parameter, l. l should be small compared to λ, the length of the diagonal of the smallest window containing all of the pattern points.

avoided, since the learning would otherwise be jerky and the result uncertain.*

When we apply the procedure defined by Figure 5.10 to a single pattern we obtain one of three possible situations:

(a) M and T are equal:
The locate, **B**, and size, F, remain unchanged.

(b) $M = +1$, $T = -1$ (**X** lies inside the hypersphere, but the teacher indicates that it should be outside). The hypersphere is moved away from **X**, and its size is reduced (Figure 5.12A).

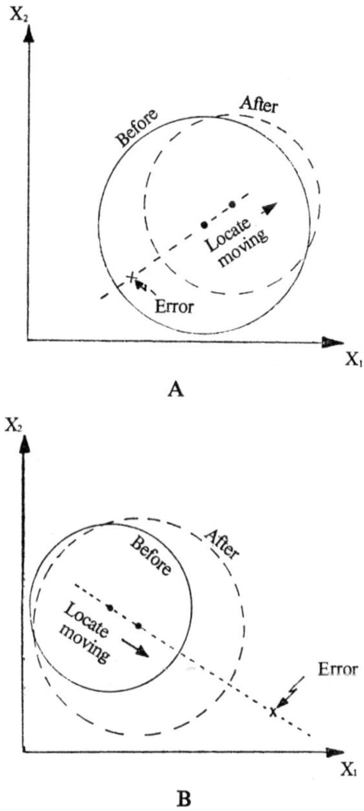

Figure 5.12 Effect of a single error on a hypersphere classifier that is being designed by the procedure of Figure 5.10. (A) $M = +1$; the error point lies inside the hypersphere before modification. (B) $M = -1$; the error point lies outside the hypersphere. In both situations, we have exaggerated the modification. Normally the adjustments would be very small, by making k and l small.

* After a long period of learning the hypersphere reaches a state of dynamic equilibrium. The hypersphere will then oscillate about the optimal solution, with random variations of its size and position. The magnitudes of this oscillation are determined by l and k. With the values suggested we have never found this random oscillation troublesome.

(c) $M = -1$, $T = +1$ (**X** lies outside the hypersphere, but the teacher indicates that it should be inside). The hypersphere is moved towards **X**, and its size is increased (Figure 5.12B).

When this procedure is applied to the complete training set the hypersphere eventually reaches a state of equilibrium (Figure 5.13). In all of

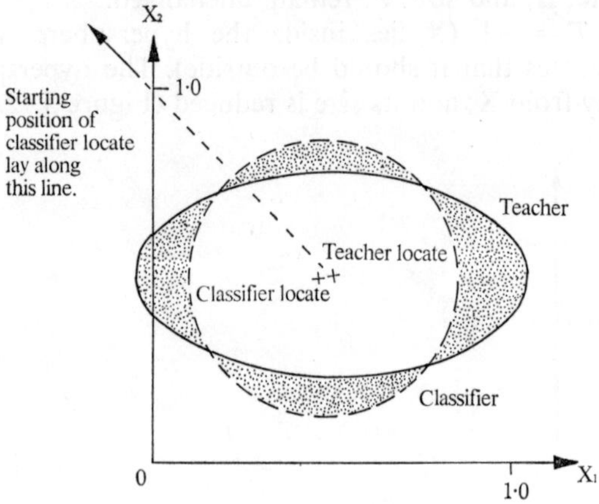

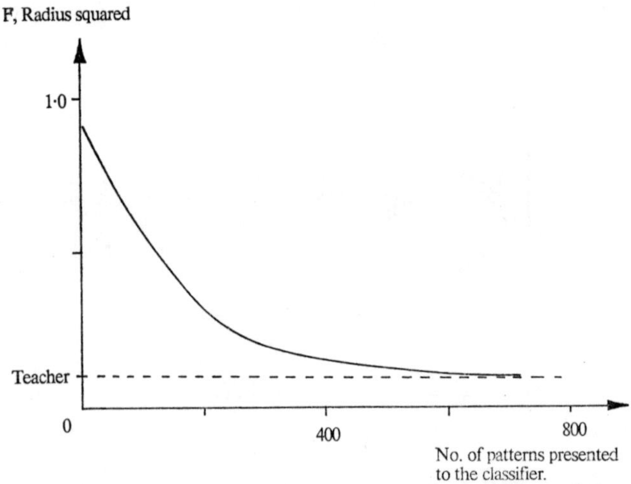

Figure 5.13 Showing the performance of the learning procedure defined in Figure 5.10. (A) Comparison of the decision surfaces of the teacher and learning classifier. Errors are indicated by stippling. ($k = 0.01$, $l = 0.01$; MANY = 5000) (B) Shows how the size parameter, F (radius squared), varies during learning. The teacher and learning classifier both used a circular decision surface. Initially their locates were coincident, but their size parameters were different. ($k = 0.001$, $l = 0.01$) (C) Locus of the locate in a circle classifier during learning. ($k = 0.01$, $l = 0.01$) (SOURCE: [7]).

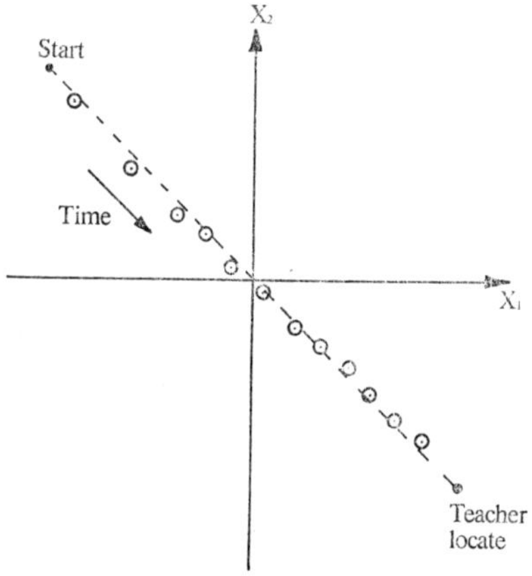

Figure 5.13 (*continued*)

our experimental work we have found a close fit between this equilibrium state and the teacher's decision surface [11], provided that the latter encloses a single compact cluster ($T = +1$ inside). The same technique can be used to train other classifiers (Figure 5.14) [12]. Of course, it would be rather eccentric to use the more complex shapes, shown in this figure, but (a) and (b) are old friends (Figure 3.17). Results obtained using these oddly shaped decision surfaces are shown in Figures 5.15 and 5.16.

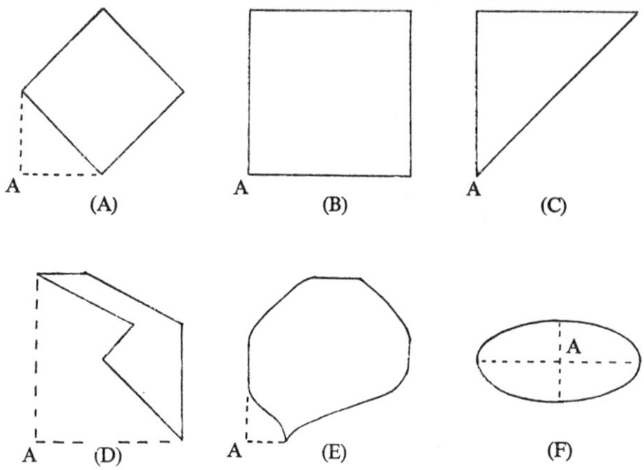

Figure 5.14 Showing the oddly shaped decision surfaces that have been trained using error-correction rules, originally developed for the hypersphere classifier. Notice how the position reference point, *A*, is defined in each case.

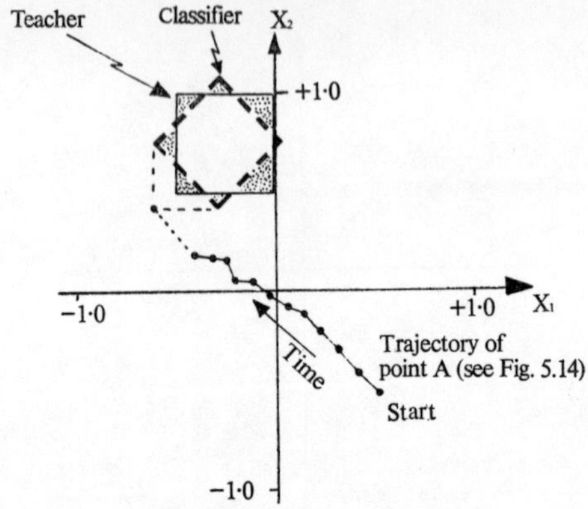

Figure 5.15 Results of training a classifier which uses the decision surface shown in Figure 5.14A (Type A experiment. $k = 0.1$; $l = 0.05$; MANY = 5000. SOURCE [7].)

In the previous section, we described a simplified technique for training a linear classifier that would be considerably simpler to implement than the original procedure. We can do the same for the hypersphere training rule. Consider Box A in Figure 5.10. The following alternative would be easier to implement:

$$\left.\begin{aligned} \mathbf{B} &:= \mathbf{B} + \{kM\text{sgn}(B_i - X_i)\} \\ \text{where} \quad \mathbf{B} &= (B_1, \ldots, B_q) \end{aligned}\right\} \tag{5.6}$$

This rule also produces satisfactory results (Figure 5.17).

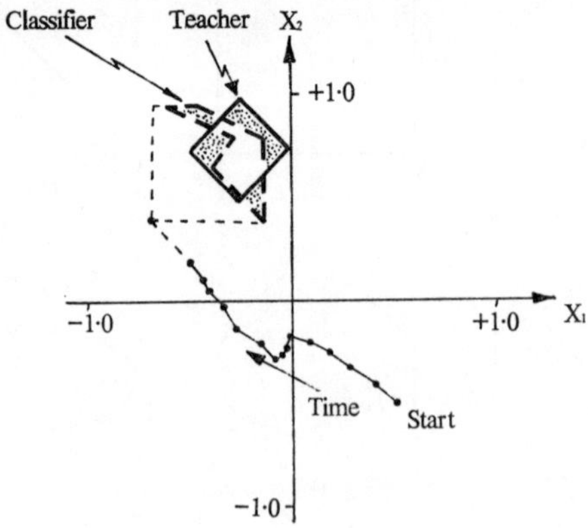

Figure 5.16 Results of training a classifier which uses the decision surface shown in Figure 5.14D. (Type A experiment. $k = 0.1$; $l = 0.05$; MANY = 5000. SOURCE [7].)

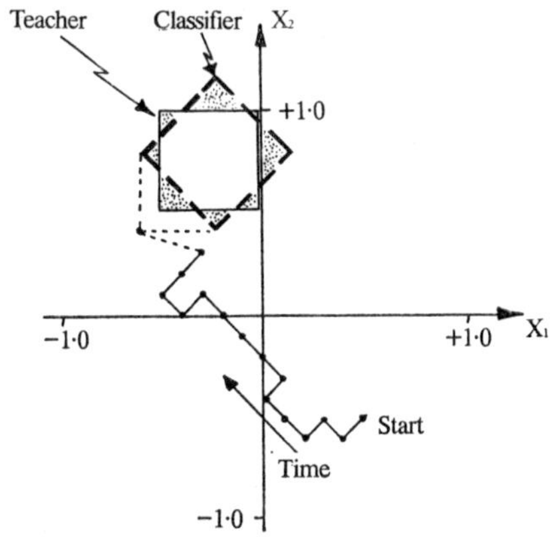

Figure 5.17 Result of training a classifier which uses the decision surface shown in Figure 5.14A. The training rule used here was the simplified rule detailed in equation (5.16). Details are otherwise exactly as given for Figure 5.15. (SOURCE [11].)

5.7 ERROR-CORRECTION LEARNING IN A CC

In the previous section we discussed techniques for training a hypersphere subclassifier. Similar methods are used for the CC, although we must add further rules for choosing which hypersphere we should move.

The flow-chart describing an error-correction procedure for training a CC is shown in Figure 5.18. Four situations occur when this is applied to a single pattern (Figure 5.19):

(a) M and T are equal.
No modifications are made.

(b) $M = -1$, $T = +1$ (**X** lies outside all of the hyperspheres). We first find that hypersphere closest to **X**. To do this, we find the value of i for which the following quantity is a minimum:

$$\sum_{j=1}^{q} (B_{i,j} - X_j)^2 - F_i \tag{5.7}$$

The ith hypersphere is moved towards **X** and its size is increased (see MODIFY 2 in Figure 5.18D, and Figure 5.20).

(c) $M = +1$, $T = -1$ (**X** lies inside *one* of the hyperspheres). The hypersphere enclosing **X** is moved away from it, and its size is reduced (see MODIFY 1 in Figure 5.18C, and Figure 5.21).

(d) $M = +1$, $T = -1$ (**X** lies inside more than one of the hyperspheres).
All hyperspheres enclosing **X** are moved away from it, and their sizes are reduced (see MODIFY 1 in Figure 5.18C, and Figure 5.22).

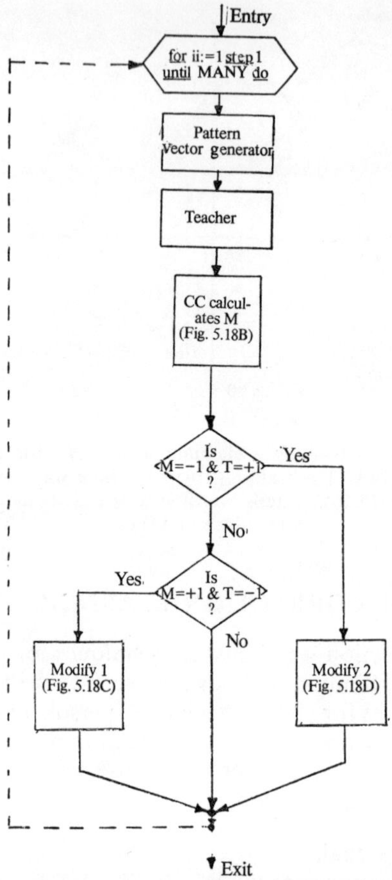

Figure 5.18 Procedure for designing a CC by error-correction learning. The notation developed in Chapter 3 has been simplified here by omitting the first suffix indicating the class. Locates are denoted by $\mathbf{B}_i$, and size parameters by F_i. (A) Iterative learning scheme in broad detail. (B) Classification. (C) First modification rule: MODIFY 1. (D) Second modification rule: MODIFY 2.

Figures 5.23 and 5.24 show typical results that we have obtained using this procedure. Further results are described in Chapter 7. See also [7, 11–14].

5.8 SUMMARY OF THE PERFORMANCE OF THE CC LEARNING RULES DEFINED IN FIGURE 5.18

(a) The procedure yields an improved CC,* although the initial choice of locates and sizes greatly influences its final performance.

* This may be judged in either of two ways: error rate, or the subjective comparison of the decision surfaces in 2-space.

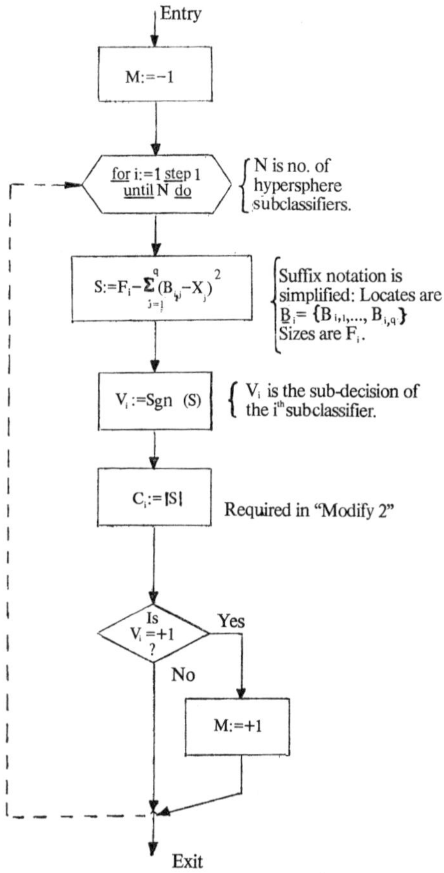

Figure 5.18 (*continued*)

(b) One (or more) of the hyperspheres will sometimes become vanishingly small during learning (degenerates). Degeneration will occur if a hypersphere were initially placed a long way from all of those points in **X**-space with $T = +1$. A degenerate hypersphere serves no useful purpose, and could be removed with no loss of accuracy. The locates labelled D in Figure 5.23B are degenerate.

(c) It is possible (though rare) for the hyperspheres to overlap to such an extent that one of them is contained (or nearly so) within the others. This is also wasteful of computing equipment.

(d) We do not yet possess a reliable method of estimating the number of hyperspheres required prior to learning [5, 15]. If we provide too many, then the classifier is unnecessarily complex, while if we provide too few, the performance will be sub-optimal.

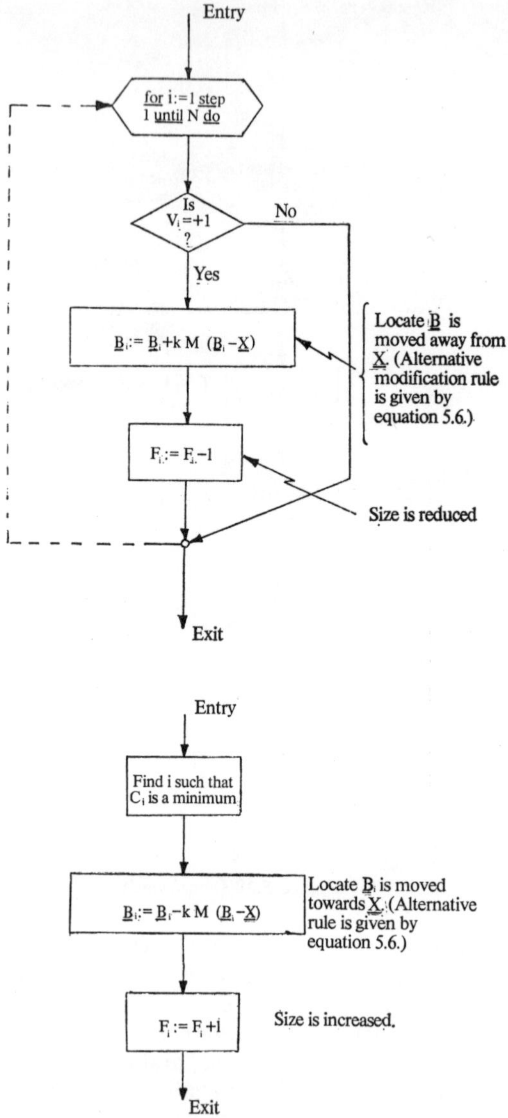

Figure 5.18 (*continued*)

(e) We have no technique for deciding, before we start training, which class should be contained within the hyperspheres. Consider Figure 5.25. It would clearly be better to associate the crosses with the teacher decision $T = +1$. If we were to invert the classes, the CC would require many more subclassifiers.

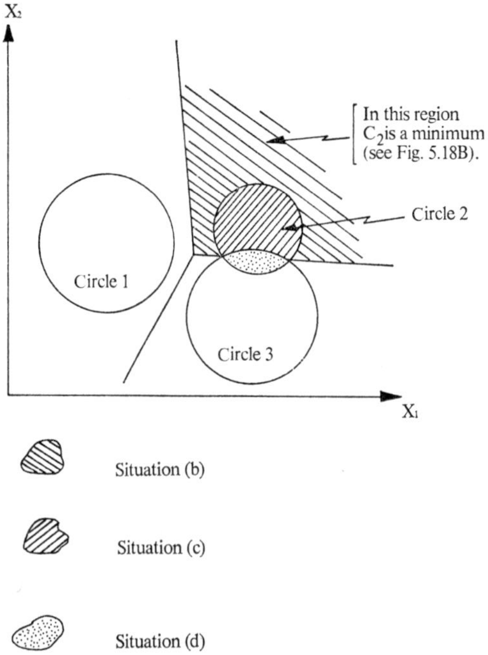

Figure 5.19 Situations encountered during learning in a CC. Errors falling within the shaded areas have the following effects:

Situation b: When $M = -1$ and $T = +1$ the procedure MODIFY 2 is applied. This leads to an increase in the size of the closest circle (circle 2) and a movement of its centre (locate) towards the mis-classified pattern.

Situation c: Here $M = +1$ and $T = -1$. The procedure MODIFY 1 is applied. This leads to a decrease in the size of circle 2, which is also moved away from the mis-classified pattern.

Situation d: Again $M = +1$ and $T = -1$, but both circles 2 and 3 enclose the mis-classified pattern. The procedure MODIFY 1 is applied. Circles 2 and 3 are both reduced in size and are moved away from the mis-classified pattern.

Situation a (not shown): When a point has been correctly classified, no modifications are made.

5.9 ERROR-CORRECTION LEARNING IN A NNC OR PFC

In view of the close relationship between the CC and NNC, we felt that we should be able to modify the procedure discussed in section 5.7 for training a NNC (16). We found this was possible, although similar difficulties to those listed in section 5.8 were encountered. At a later date the similarity between the NNC and the heteropolar PFC became apparent, and we then applied the same learning procedure to a number of PFCs (17). So far, we have not attempted to train a homopolar PFC using this technique.

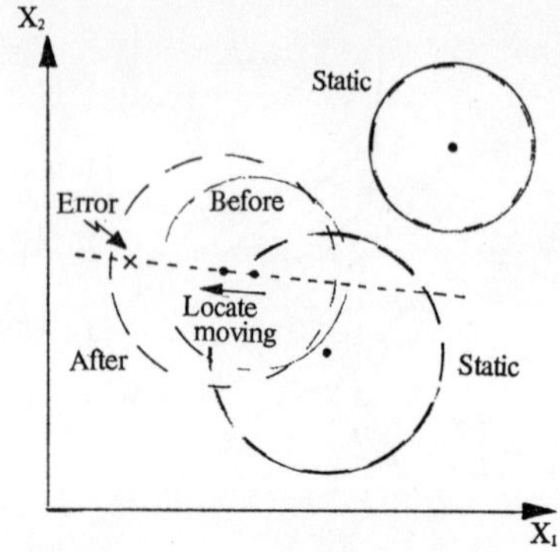

Figure 5.20 Effect of a single error, typical of situation b, when designing a CC by error-correction learning.

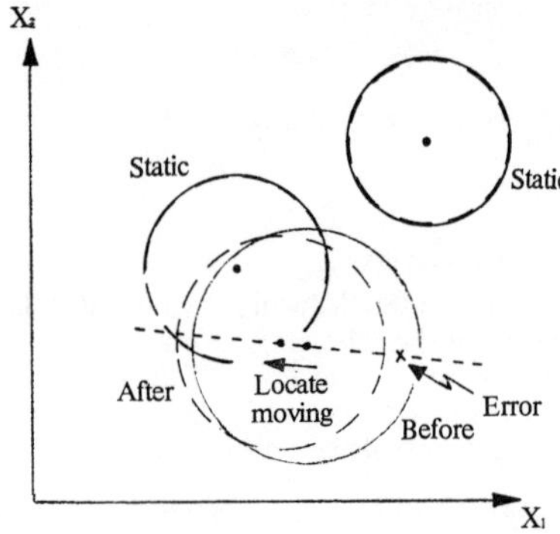

Figure 5.21 Effect of a single error, typical of situation c, when designing a CC by error-correction learning.

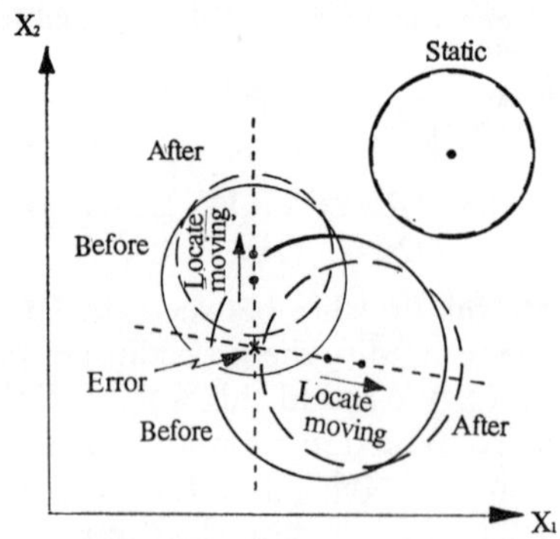

Figure 5.22 Effect of a single error, typical of situation d, when designing a CC by error-correction learning.

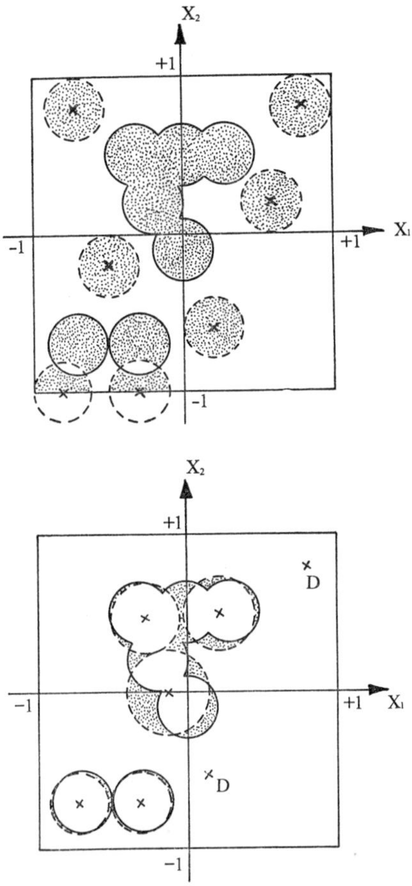

Figure 5.23 Results of training a CC by the procedure defined in Figure 5.18. The square in (A) and (B) indicates the extreme ranges of the variables X_1 and X_2. Errors are indicated by stippling. (A) Initial classifier (broken) compared to the teacher (solid). (B) Classifier, after learning, compared to the teacher. Points indicated by D are the positions of degenerate circles (zero area). (C) Learning curve. (Type A experiment. $k = 0.02$; $l = 0.0015$; MANY = 5000. SOURCE [12].)

The CC learning procedure cannot be transferred directly to the NNC, since the former is a homopolar device, while the latter is its heteropolar equivalent. Nevertheless, the modifications are fairly obvious (Figure 5.26). When we are training a NNC or PFC we modify two locates. We select one locate from each class for modification, by considering the weighted distances between the point producing the error and the locates (Figure 5.26, Box A). The chosen locates are then moved directly towards, or away from, the error, and the corresponding weights are adjusted by a simple decrement/increment rule.

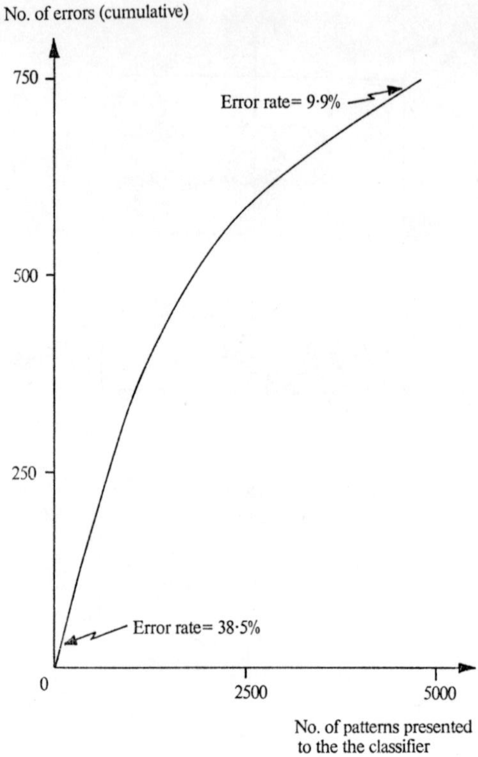

Figure 5.23 (*continued*)

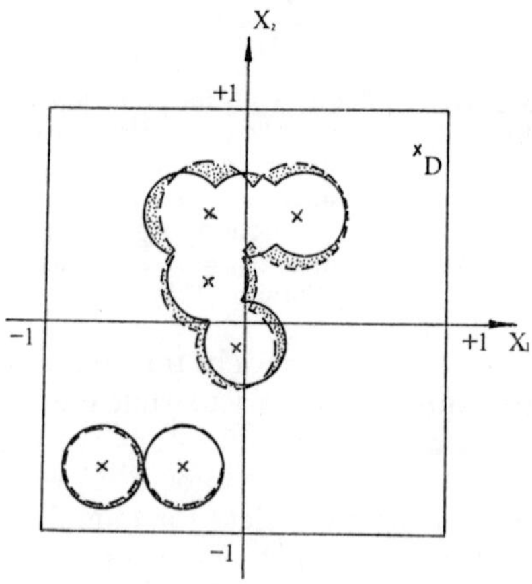

Figure 5.24 As Figure 5.23, except that the simplified locate modification rule, described in section 5.6, equation (5.6), was used. (A) Final classifier compared to the teacher. (B) Learning curve. (SOURCE [12].)

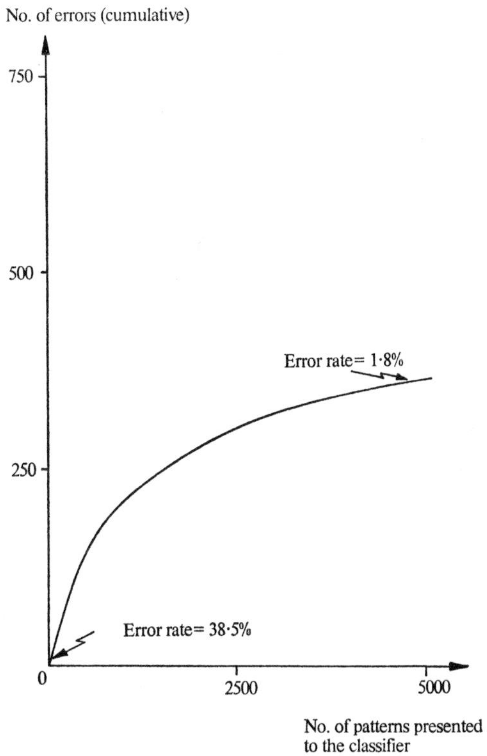

Figure 5.24 *(continued)*

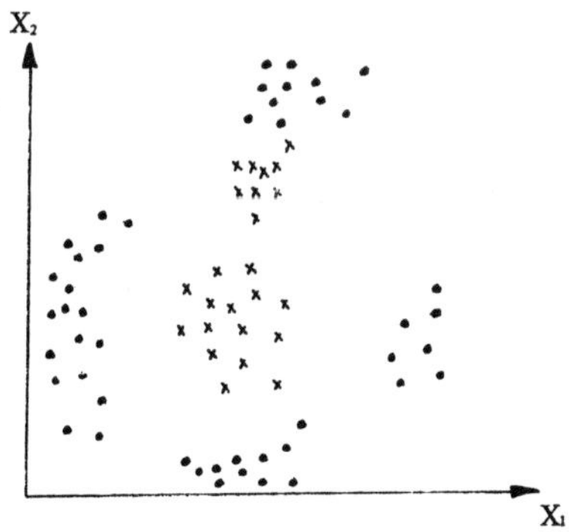

Figure 5.25 It would be preferable (using fewer circles) to associate the crosses with the teacher decision $T = +1$. If we do this, we should require only one circle in a CC. If we associate black circles with this teacher response, we should need at least three circles in our CC.

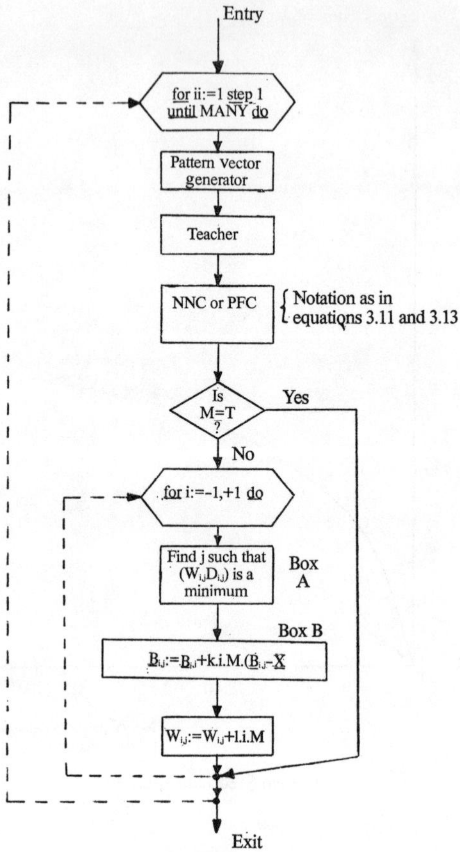

Figure 5.26 Procedure for designing a NNC or PFC by error-correction learning.

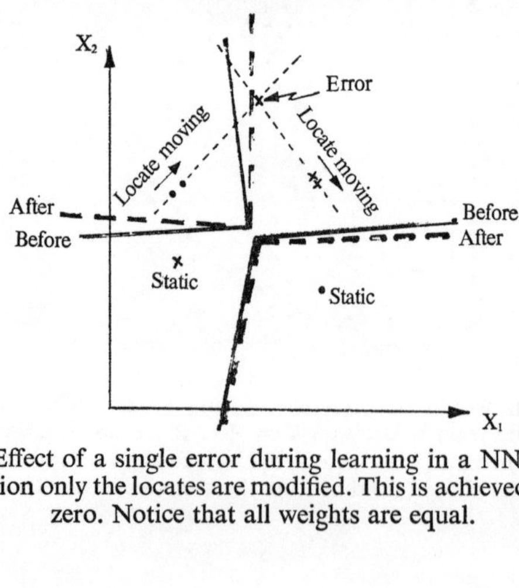

Figure 5.27 Effect of a single error during learning in a NNC. For clarity, in this illustration only the locates are modified. This is achieved by setting *l* to zero. Notice that all weights are equal.

Figure 5.27 shows the way a simple NNC responds to a single error. Notice that in this figure the iteration parameter, l, is zero. Typical values of the iteration parameters are:

$$\begin{aligned} 0.0001 \leq k \leq 0.05 \\ 0 \leq l \leq 0.01\ W \end{aligned} \tag{5.8}$$

where W is the average value of the weights.

Typical results obtained with this learning procedure are shown in

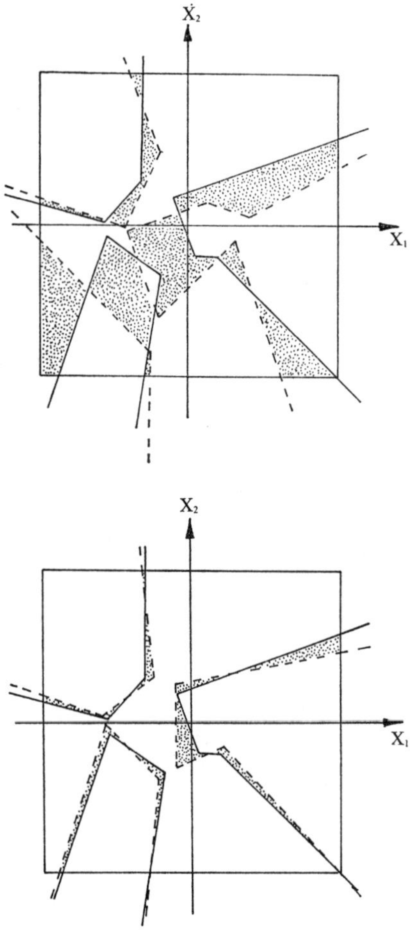

Figure 5.28 Result of training a NNC using the procedure defined in Figure 5.26. The teacher, and learning classifier both used 8 locates. (A) Initial classifier (broken) compared to the teacher (solid). (B) Classifier, after learning, compared to the teacher. (Type A experiment. k was progressively reduced from 0.05 to 0.005; $l = 0.0$; MANY = 3100. SOURCE [16].)

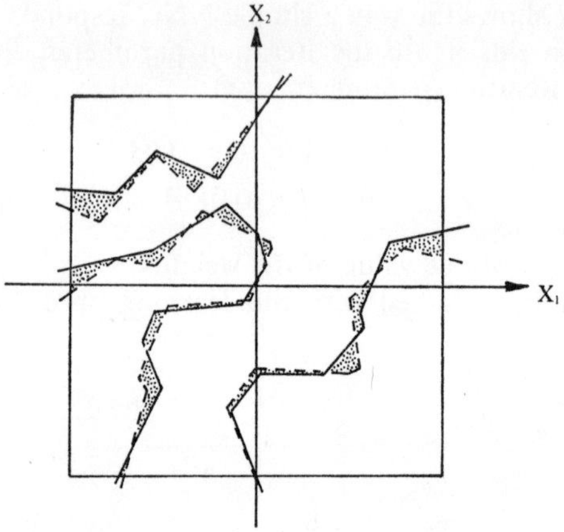

Figure 5.29 Result of training a NNC using the procedure defined in Figure 5.26. (Type A experiment. $k = 0.02$ for 3000 iterations, followed by 0.01 for 2000 iterations; $l = 0.0$. SOURCE [16].)

Figures 5.28–5.29. From these and many other similar results we reached the following conclusions, which closely match those listed in section 5.8:

(a) The procedure yields an improved classifier, although the initial choice of the locates and weights greatly influences the results.
(b) If one of the locates is placed a long way from patterns belonging to its own class, the corresponding weight tends to become very large. In this case the "island" surrounding that locate degenerates and becomes vanishingly small. This is wasteful of computing equipment.
(c) It is possible for the island surrounding one locate to become completely embedded (or nearly so) within islands of the same class. The locate giving rise to the smaller island could be removed with no deterioration of the performance of the classifier (Figure 5.30).
(d) There is no reliable method for estimating the number of locates required prior to learning [13].

There is, however, one additional difficulty which arises with the learning procedure defined in Figure 5.26. The locates tend to move together during learning, and may eventually become so close that instability occurs (section 4.7, type 1 instability). The problem may be overcome by using a different locate modification rule (Box B, Figure 5.26). For example, instability may be avoided by ensuring that both modified locates move through equal distances. This can be accomplished by the following locate modification rule:

$$\mathbf{B}_{i,j} := \mathbf{B}_{i,j} - kiM \frac{(\mathbf{B}_{i,j} - \mathbf{X})}{|\mathbf{B}_{i,j} - \mathbf{X}|} \qquad (5.9)$$

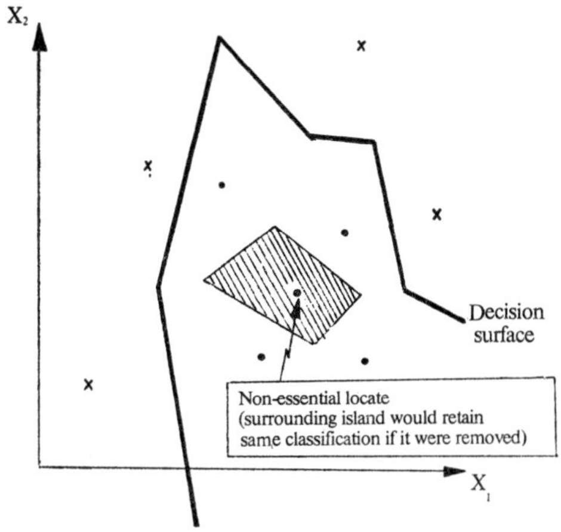

Figure 5.30 Redundant locate in a NNC.

5.10 GROWING AND PRUNING

The major objection to the learning techniques described in sections 5.7 and 5.9 is that they do not indicate whether the classifier contains enough locates. This would not be so important if we could reliably anticipate the number required, prior to learning. Some research workers have proposed the use of cluster analysis techniques to do this. However, these place greatest emphasis on the peaks of the pdfs, whereas the tails are more important in decision-making. For this reason, we should not overstate the value of mode-seeking procedures (Appendix 2).

We shall describe how parameter adjustment learning may be combined with the addition or removal locates to form much more powerful techniques, which we shall call *Growing* and *Pruning* [7, 13, 14, 18]. These employ the parameter-adjustment learning rules in two ways:

(a) To improve the performance of a classifier as we have already described in sections 5.7 and 5.9.
(b) To provide a method for locating which regions of the descriptor space yield the greatest number of errors.

The second of these may be regarded as a technique for *scoring* the locates. Those with high scores are then removed (pruning), or "split" to form two new locates (growing).

One difficulty remains: how do we decide whether we should grow or prune at any given instant? We shall return to this problem in Chapter 6, where we discuss the practical aspects of classifier design. For the moment, we shall assume that there is no difficulty in deciding whether to grow or prune our classifier.

5.11 GROWING IN A CC

Growing is a procedure by which the complexity of a pattern classifier may be increased after it has been found that further parameter adjustment learning cannot yield an improved performance. Growing may be likened to the procedure followed by a statistician who wishes to fit a polynomial curve to a set of data. He would probably start by attempting to fit a linear function, using regression techniques. He would then test the goodness of fit, finding that he required a higher order function. The next stage is the use of regression methods to fit a quadratic function. This is followed by testing, and possibly increasing the complexity of the polynomial function to a cubic. The statistician goes through the following steps:

(1) Hypothesise that a simple polynomial function (linear) will fit the data.
(2) Optimise the coefficients associated with the current polynomial, using regression.
(3) Test whether the current polynomial fits the data adequately. If it does, then the procedure is terminated. If not, steps (4) and (5) follow.
(4) Increase the order of the current polynomial by one.
(5) Repeat, by returning to (2).

The design of a CC by growing follows this approach very closely:

(i) Start the classifier with one locate.
(ii) Optimise the positions and sizes of the hyperspheres, using parameter-adjustment learning.
(iii) Test whether the classifier performance is satisfactory. (Is the error rate low enough?) If it is, then the procedure is terminated. If not, (iv) and (v) follow.
(iv) Add another hypersphere subclassifier.
(v) Repeat, by returning to (ii).

The signal flow-chart which defines a procedure for growing a CC,* is shown in Figure 5.31. We shall describe the performance of this procedure with reference to Figure 5.32. (This diagram presents one of our experimental results. Notice that the city block distance is used. In all other respects the method used to obtain these results was identical with Figure 5.31.)

(a) Figure 5.32D shows the decision surface used by the teacher.†
(b) Figure 5.32A shows the decision surface of the CC before any learning (broken curve). Just one subclassifier is used ($N = 1$ in

* Other techniques have been suggested for growing a CC [7, 18]. The procedure discussed here is the one that has achieved the best results.

† See legend to this figure.

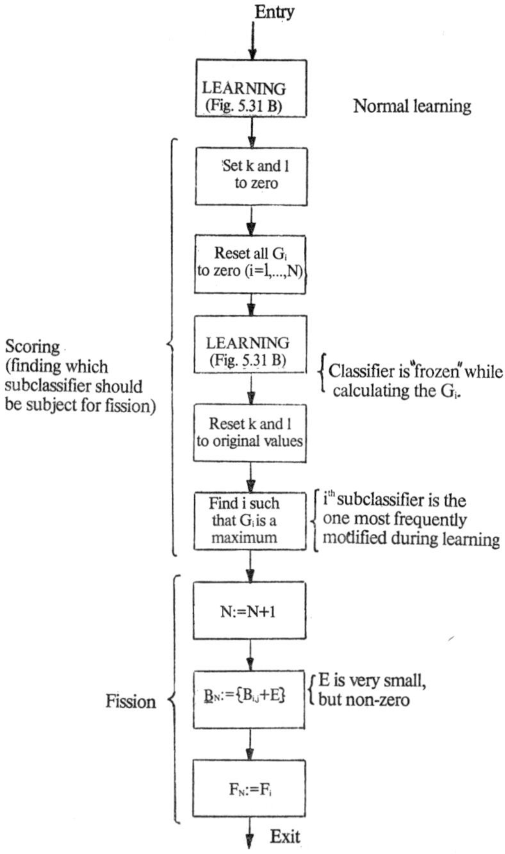

Figure 5.31 Procedure for growing a CC by fission. (A) Growing, in broad detail. (B) Iterative learning scheme, plus calculation of scores (*cf.* Figure 5.18A).

Figure 5.31). This figure also shows the decision surface of the CC after parameter-adjustment learning (solid curve).

(c) We now add another square, very close to the solid square in Figure 5.32A and with the same size. We have called this process *fission* (Figure 5.31). The new square is added close to the original, but not coincident with it. This can be guaranteed if we make the quantity E (Figure 5.31) small, compared to (F_1/q), but still finite. By adding the second subclassifier, we have increased the *potential* flexibility of the CC without making any significant change to its decision surface.

(d) We now apply the parameter adjustment learning rules to the enlarged CC ($N = 2$). The resulting decision surface is shown in Figure 5.32B.

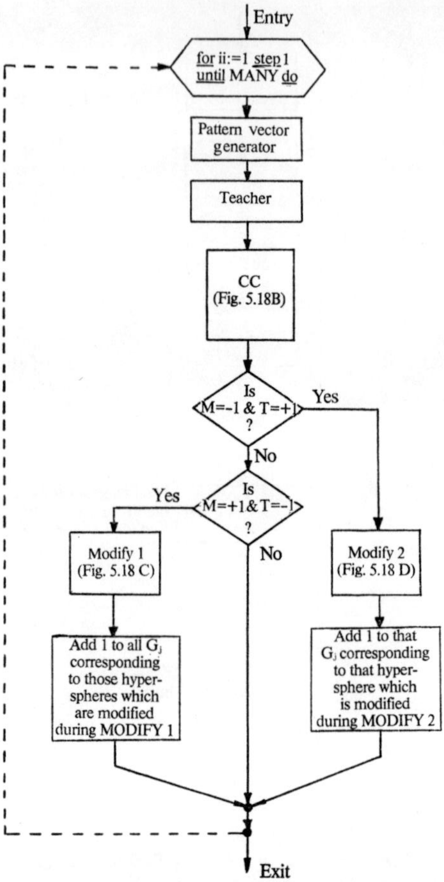

Figure 5.31 (*continued*)

(e) The upper square in Figure 5.32B, labelled NEXT, is now selected for fission. The method of selection is described below (also see Figure 5.31).

(i) We temporily set k and l to zero. This allows us to operate the parameter-adjustment learning procedure to calculate the locate scores, without disturbing the decision surface. If, as we expect, the learning has reached a stationary state, there is strictly no need to "freeze" the classifier like this.

(ii) We now apply the parameter-adjustment learning procedure to calculate the growing scores (G_j in Figure 5.31). G_j records the probability that the jth subclassifier would be modified, if learning were allowed to continue. Hence G_j is a measure of the mismatch between the CC and teacher surfaces in the immediate neighbourhood of the jth square.

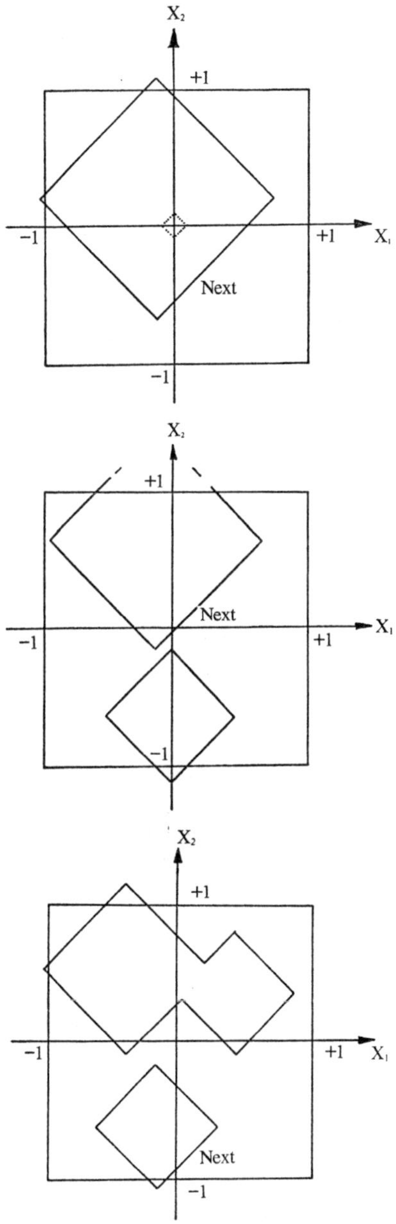

Figure 5.32 A simple example of growing in a CC which uses the city block distance. For an explanation of the principles of growing see the text. (A) Initial CC decision surface (broken) and after learning (solid). (B) Decision surface of the growing CC after one stage of fission, plus learning. (C) Decision surface of the growing CC after two stages of fission, plus learning. (D) Teacher decision surface. After 3 stages of fission plus learning, the decision surface of the growing classifier was such a good fit that no errors were observed in a test set of 2500 patterns. (Type A experiment. $k = 0.05$; $l = 0.01$; MANY = 10 000. SOURCE [14].)

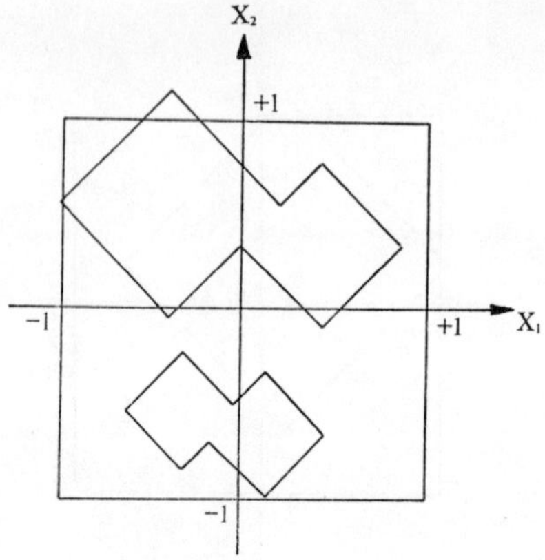

Figure 5.32 (*continued*)

(A better example is seen in Figure 5.32C where the bottom square has a large value for G_j, because it cannot match the two teacher clusters adequately.)

(iii) Let us turn our attention back to Figure 5.32B. We then find that square which has the largest growing score. This is the one labelled NEXT, and will be the subject of the fission process.

(f) We continue to apply the learning, scoring and fission procedures in rotation. Figure 5.32C shows the CC surface with 3 squares, while Figure 5.32D shows the surface with 4 squares. (This is such a close fit to the teacher surface that the difference is not visible.)

Further results of growing are presented in Figures 5.33, 5.37 and 5.38.

5.12 PRUNING IN A CC

Suppose we have designed a CC which has a low error rate, but contains some redundant locates. (Such a situation might arise because we had continued to grow after the classifier had reached its optimal size.) We could reduce the size of the classifier, without worsening its performance, by simply removing the redundant locates (Figure 5.34). However, we find in practice that complete redundancy is rare, whereas nearly-redundant subclassifiers* are more common. It is preferable to

* A typical nearly-redundant hypersphere may contain one or two points whose classification would change, if it were removed.

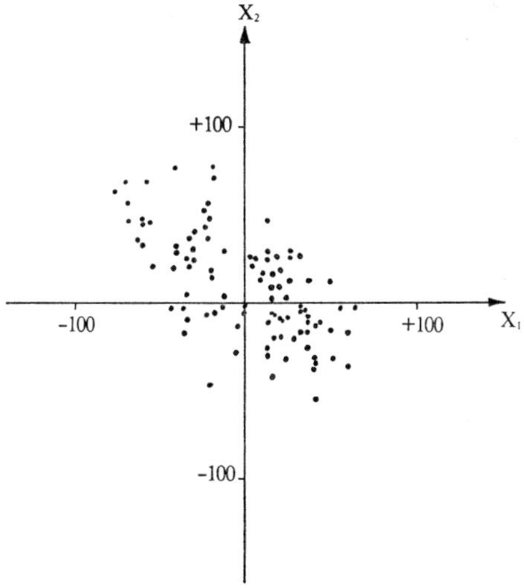

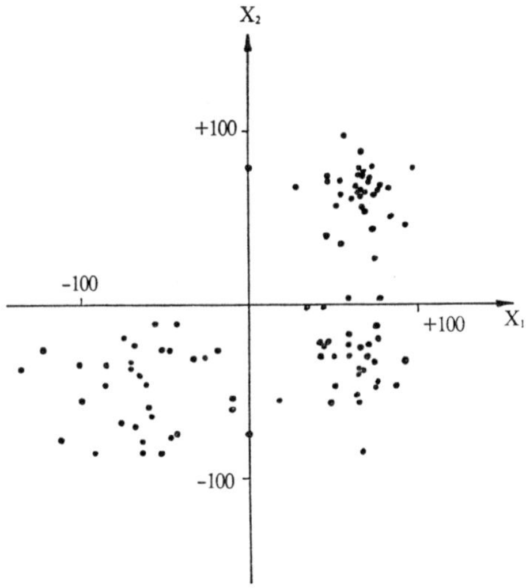

Figure 5.33 Results of growing a CC by the procedure defined in Figure 5.31. The generation and statistics of this data are discussed in Appendix 3. (A) Class corresponding to $T = -1$. (B) Class corresponding to $T = +1$. (C) Initial CC decision surface. (D) CC decision surface after learning. (E) CC decision surface soon after the first stage of fission. (F) A short period of learning after (E). (G) After further learning. The decision surface is now static. (H) After 2 stages of fission and learning the decision surface had reached this position. (Type B experiment. Data is listed in Appendix 3. $k = 0.005$; $l = 10.0$; MANY = 10 000.)

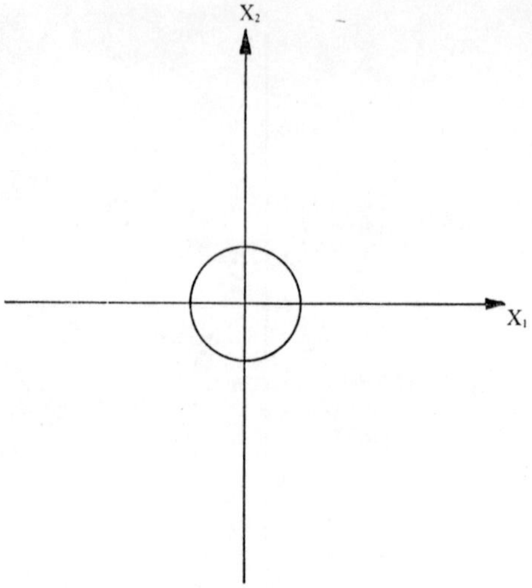

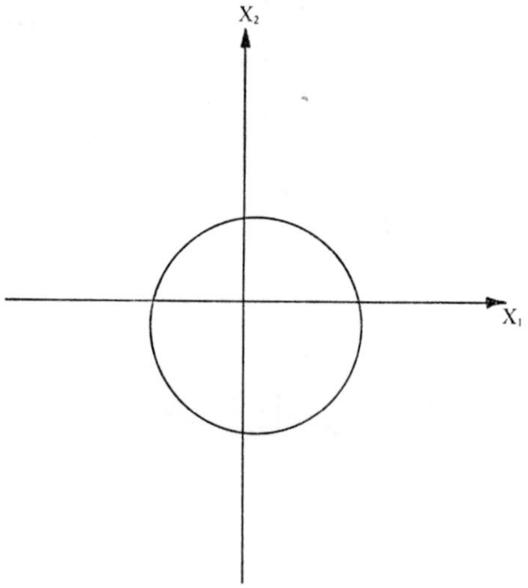

Figure 5.33 (*continued*)

regard redundancy as relative, since there is no clear division between redundant and non-redundant locates. For this reason, we developed a locate deletion procedure (pruning) which *normally* removes just one hypersphere at a time. After each reduction in classifier size, we apply the familiar parameter-adjustment rules to ensure that the error rate remains as low as possible. The pruning procedure is defined in Figure 5.35. The method of scoring (estimating redundancy) may be readily understood with reference to Figure 5.36. Points within Y contribute

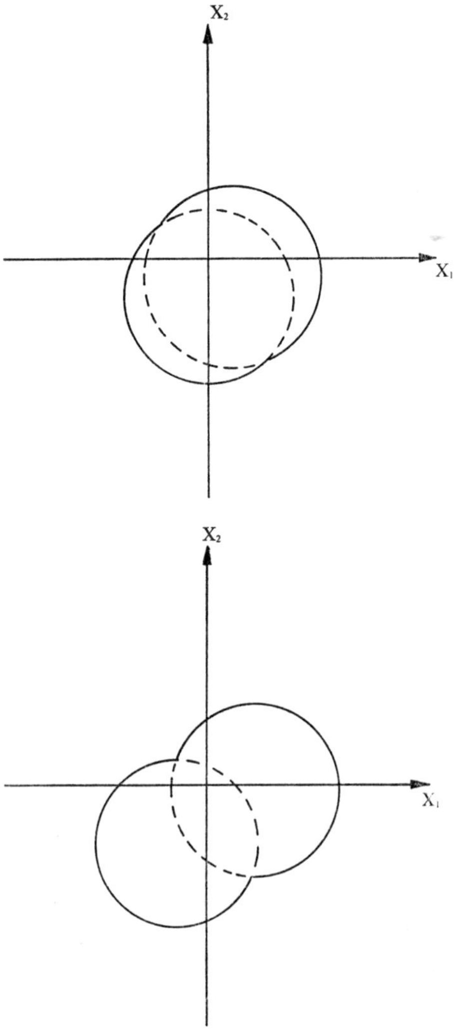

Figure 5.33 (*continued*)

to the Y_1 score, and are misclassified because circle 1 is present. Those points within Z contribute to Z_1, and are correctly classified solely as a result of the presence of circle 1. The quantity $(Y_1 - Z_1)$ therefore represents the improvement that would result, if circle 1 were removed. It is worthwhile making three points here:

(a) The decision to define Y as that region, within the circle *and* in which $T = -1$, was quite arbitrary. It might be argued that the

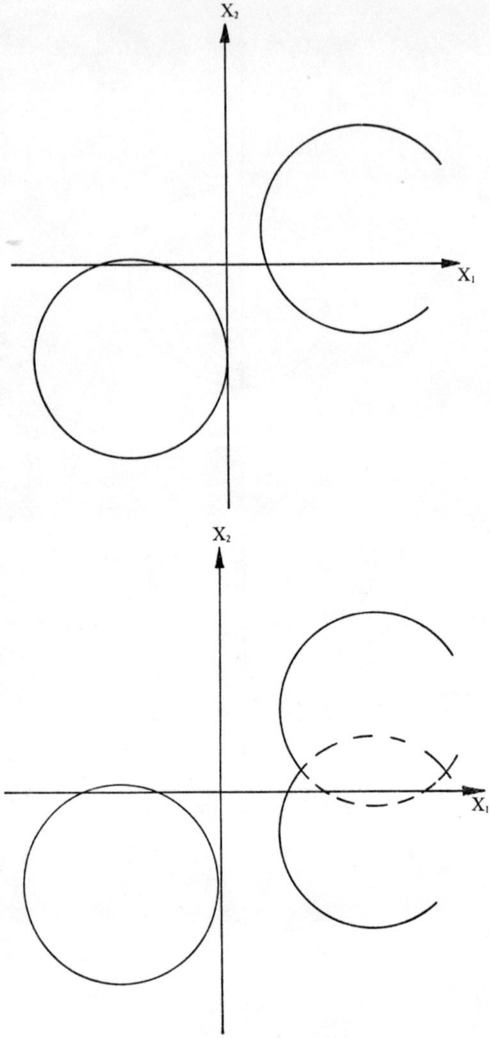

Figure 5.33 (*continued*)

following is a better definition of the area contributing to Y_1:

$$\begin{cases} \text{inside circle 1,} \\ \text{outside circle 2 (and all others),} \\ T = -1. \end{cases}$$

However, the method we use is computationally simpler, and it works satisfactorily.

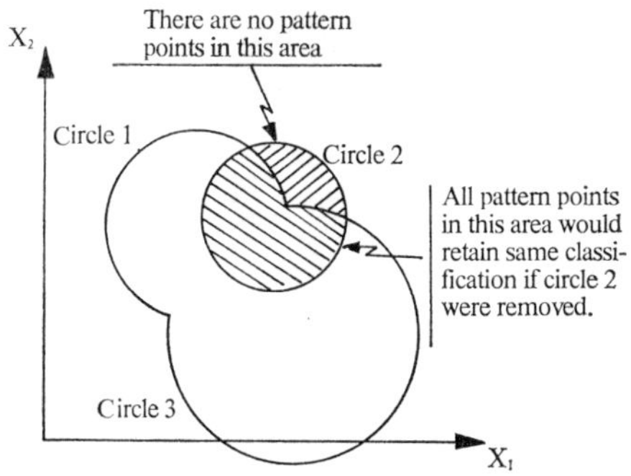

Figure 5.34 Illustrating the need for a procedure for pruning. Circle 2 is totally redundant.

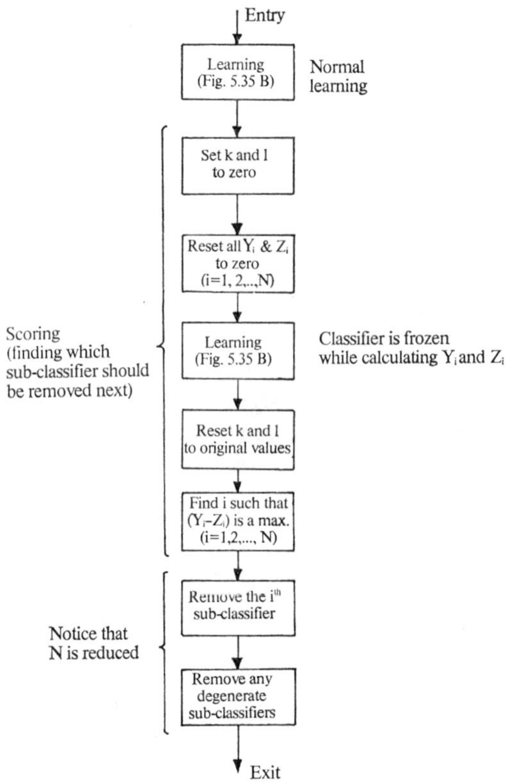

Figure 5.35 Procedure for pruning a CC. (A) Pruning, in broad detail. (B) Iterative scheme for learning and scoring (*cf.* Figures 5.18A and 5.31B). (C) Details of the scoring procedure.

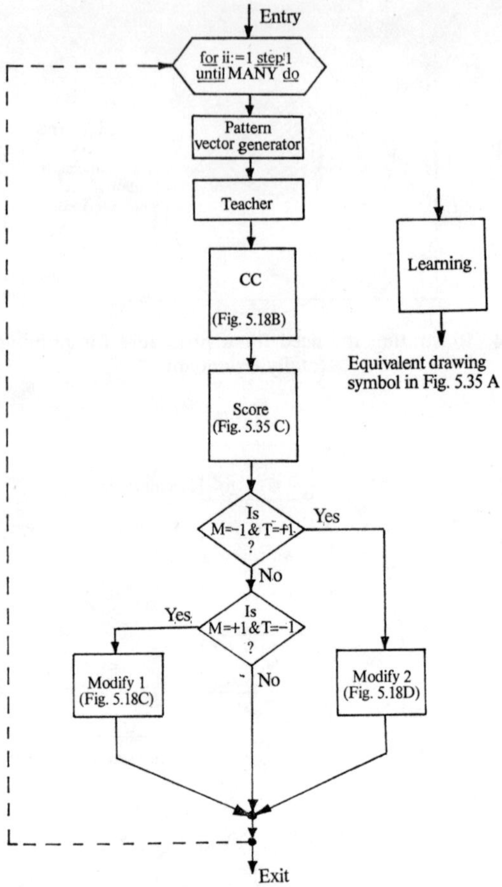

Figure 5.35 (*continued*)

(b) It is often worthwhile continuing to prune, even if the quantity

$$\max_{i=1,\ldots,N} (Y_i - Z_i) \tag{5.10}$$

is negative.

(c) The flow-chart shown in Figure 5.35 also includes a provision for deleting any degenerate subclassifiers (section 5.8).

Some of our typical results, obtained using these pruning rules, are shown in Figures 5.37 and 5.38.

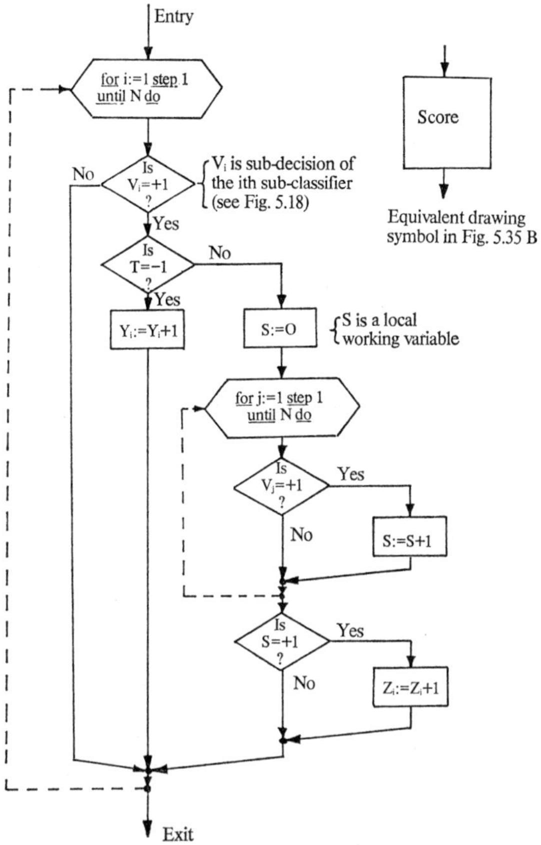

Figure 5.35 (*continued*)

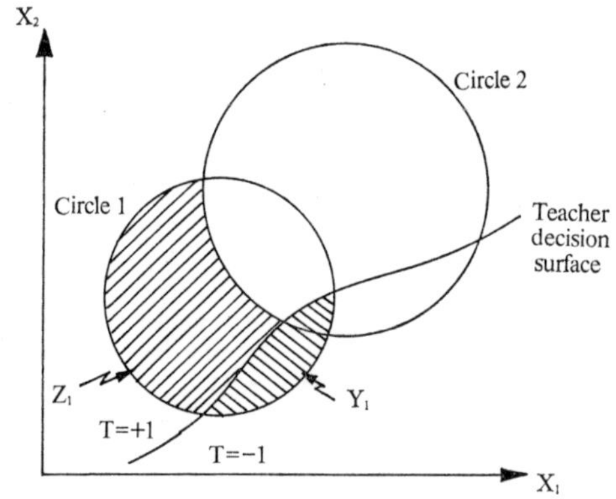

Figure 5.36 Showing the significance of the pruning score, which is equal to $(Y_1 - Z_1)$.

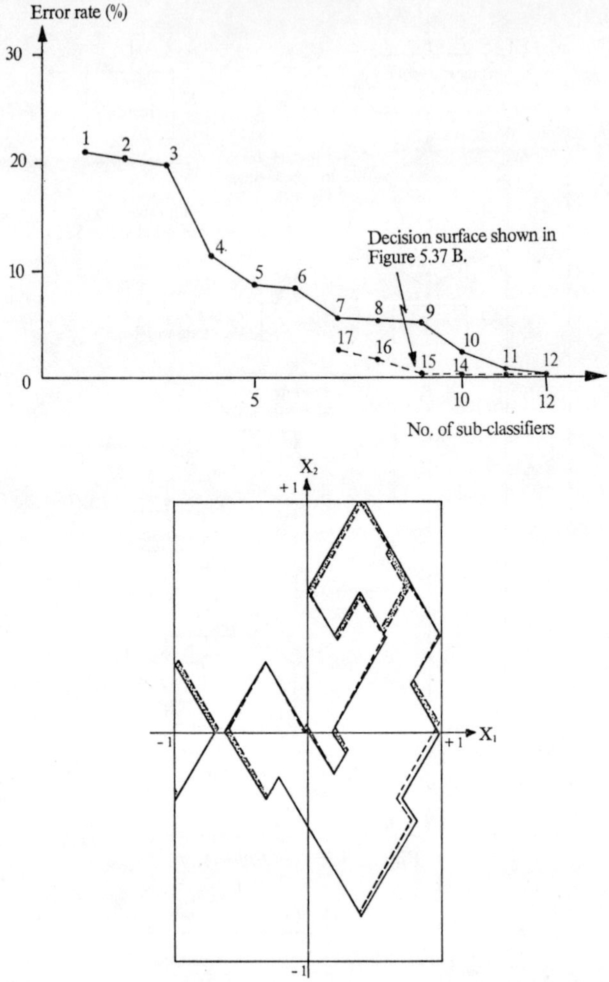

Figure 5.37 Results of designing a CC by growing and pruning. The teacher and adaptive classifier both used the city block distance. The teacher used 9 locates. (A) Error rate plotted against classifier size. Numbers indicate the sequence of generating the various CCs. The growing rules were applied until there were 12 subclassifiers, after which pruning was used. The "best" classifier, in terms of size and error rate, is number 15. (B) Decision surface of the CC with sequence number 15 (broken curve) compared to that of the teacher (solid). Errors are indicated by stippling. This figure was obtained by tracing the output from a computer. Hence, the axes, X_1 and X_2, have been rescaled in the ratio 3:5. (Type A experiment. $k = 0.05$; $l = 0.01$; MANY = 10 000. Source [14].)

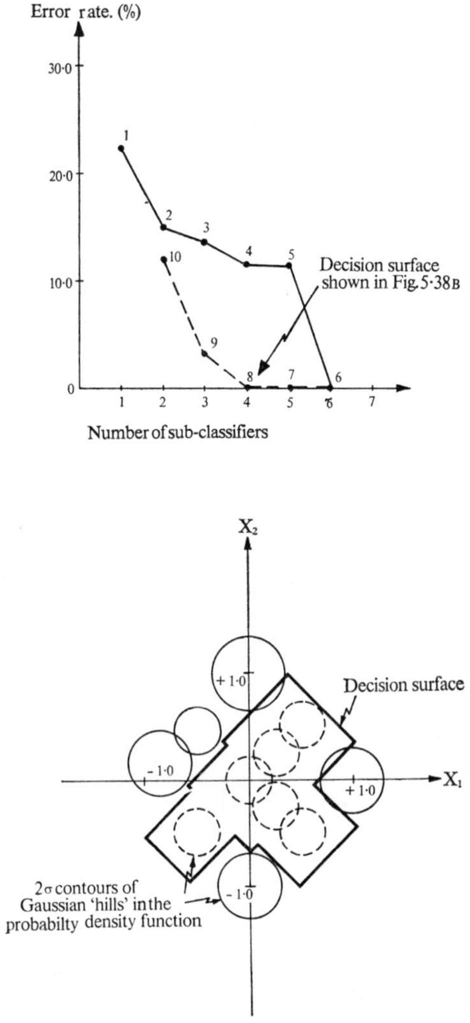

(B)

Figure 5.38 Results of designing a CC by growing and pruning. (A) Error rate plotted against classifier size. Numbers indicate the sequence of generating the various CCs. The growing rules were applied until there were 6 subclassifiers, after which pruning was used. The "best" classifier in terms of size and error rate is number 8. (B) Decision surface of the CC with sequence number 8 compared to the contours of the probability density functions. (Type B experiment. $k = 0.05$; $l = 0.025$; MANY = 5000. Source [14].)

5.13 LEARNING, GROWING AND PRUNING IN A MULTI-CATEGORY NNC OR PFC

Unfortunately, we must change our notation to describe training rules for a multi-category classifier. We have previously used a double subscript notation for locates and weights, whereas in this section, we shall use a single subscript. Hence, the locates will be denoted by $\mathbf{B}_1$, $\mathbf{B}_2, \ldots$ and the weights by W_1, $W_2, \ldots$. The locates will be allocated to the classes in the following way:

$$\left.\begin{array}{l} \mathbf{B}_1 \\ \cdot \\ \cdot \\ \cdot \\ \mathbf{B}_{N_1} \end{array}\right\} \text{Class 1}$$

$$\left.\begin{array}{l} \mathbf{B}_{N_1+1} \\ \cdot \\ \cdot \\ \cdot \\ \mathbf{B}_{N_1+N_2} \end{array}\right\} \text{Class 2}$$

$$\left.\begin{array}{l} \mathbf{B}_{N_1+N_2+1} \\ \cdot \\ \cdot \\ \cdot \\ \cdot \\ \mathbf{B}_{N_1+N_2+N_3} \end{array}\right\} \text{Class 3}$$

etc.

The weights are also distributed in this way. Later in this section, we shall require growing and pruning scores, which we shall include in our flow-charts from the outset. The growing scores will be denoted by G_1, $G_2, \ldots$, and the pruning scores by P_1, $P_2, \ldots$.

We shall describe learning rules which employ both error-correction and reinforcement learning. We have found reinforcement learning particularly useful in helping to maintain large interlocate distances (section 4.7.1 and 5.9). We therefore require iteration parameters for both types of learning:

$$\left.\begin{array}{l} k\text{, controls magnitude of locate movement} \\ l\text{, controls increment on weights} \end{array}\right\} \text{error correction}$$

$$\left.\begin{array}{l} k_r\text{, controls magnitude of locate movement} \\ l_r\text{, controls increment on weights} \end{array}\right\} \text{reinforcement}$$

5.13.1 Learning

We are now in a position to be able to define the learning rules (Figure 5.39), which are briefly described below:

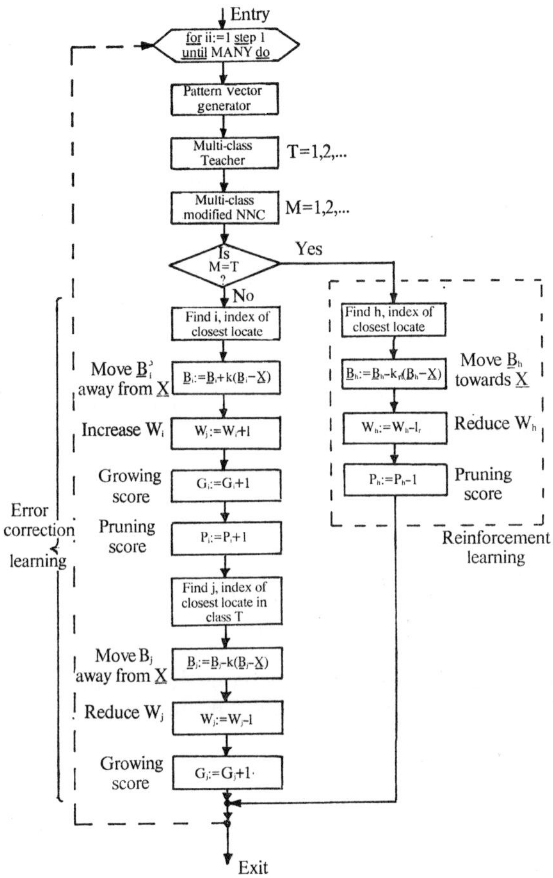

Figure 5.39 Procedure for learning in a multi-category NNC. $\mathbf{B}_1$, $\mathbf{B}_2$, . . . are locates. W_1, W_2, . . . are weights. k, l, k_r and l_r are positive constants, called iteration parameters. G_1, G_2, . . . are scoring counters, used in growing. P_1, P_2, . . . are scoring counters, used in pruning. The "closest locates" are defined in terms of weighted distances measured from the **X**-vector.

Error correction learning

Let us assume that **X** is a member of class 1 ($T = 1$), but that it has incorrectly been attributed to class 3 ($M = 3$) due to the proximity* of

* Measured in terms of weighted distances.

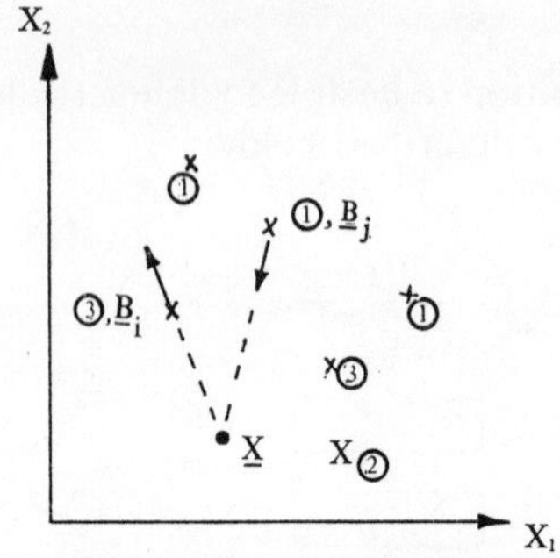

Figure 5.40 Effect of a single error in a NNC during the operation of the procedure defined in Figure 5.39. **X** is the mis-classified point. Class membership of the locates is denoted by the encircled numbers. In the following description "closeness" is measured in terms of the weighted Euclidean distance from **X**. $\mathbf{B}_j$ is the closest class 1 locate, and $\mathbf{B}_i$ the closest locate of all (class 3). The locate $\mathbf{B}_j$ is moved towards **X** and its weight W_j is reduced, while $\mathbf{B}_i$ is moved away from **X** and W_j is increased.

locate $\mathbf{B}_i$. Let $\mathbf{B}_j$ be that locate in class 1 which is closest* to **X**. The situation is shown in Figure 5.40. The locates and weights are modified as follows:

$$\begin{aligned} \mathbf{B}_j &:= \mathbf{B}_j - k(\mathbf{B}_j - \mathbf{X}), \quad && \text{moved towards } \mathbf{X}; \\ \mathbf{B}_i &:= \mathbf{B}_i + k(\mathbf{B}_i - \mathbf{X}), && \text{moved away from } \mathbf{X}; \\ W_j &:= W_j - l, && \text{reduced;} \\ W_i &:= W_i + l, && \text{increased.} \end{aligned} \tag{5.11}$$

All other locates and weights are unchanged, including any from (say) class 2 which lie intermediate* between $\mathbf{B}_i$ and $\mathbf{B}_j$.

Reinforcement learning

Let us assume that **X** was correctly assigned to class 3 by the proximity of $\mathbf{B}_h$. This locate and its corresponding weight are modified as follows:

$$\begin{aligned} \mathbf{B}_h &:= \mathbf{B}_h - k_r(\mathbf{B}_h - \mathbf{X}), \quad && \text{moved towards } \mathbf{X}; \\ W_h &:= W_h - l_r, && \text{reduced.} \end{aligned} \tag{5.12}$$

This situation is illustrated in Figure 5.41, except that we have constrained all weights to be equal, and have set l_r to zero.

5.13.2 Growing and Pruning

Growing and pruning are defined in Figures 5.42 and 5.43, respectively. These procedures closely resemble the rules for varying the size

* Measured in terms of weighted distances.

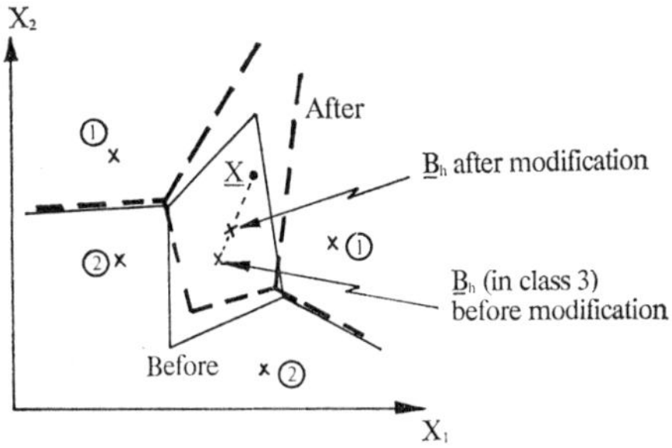

Figure 5.41 Effect of a single modification due to reinforcement learning in a multi-category NNC. $\mathbf{B}_h$ is the closest locate to **X**. ("Closeness" is measured in terms of weighted distance.) $\mathbf{B}_h$ is moved towards **X** and its weight is reduced. For clarity in this illustration, all weights are equal, and the iteration parameter l_r is zero.

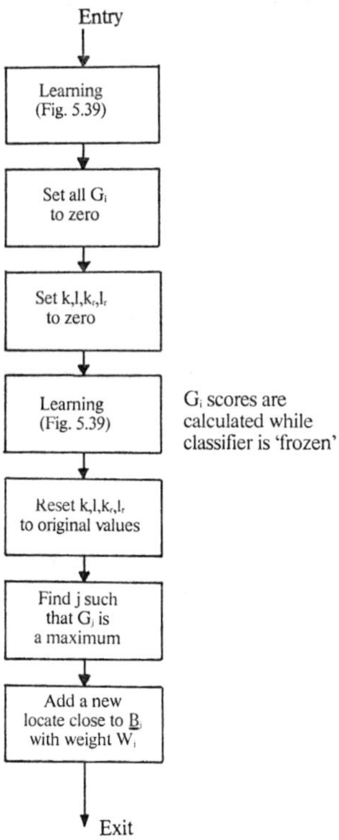

Figure 5.42 Procedure for growing a multi-category NNC. Notation as in Figure 5.39.

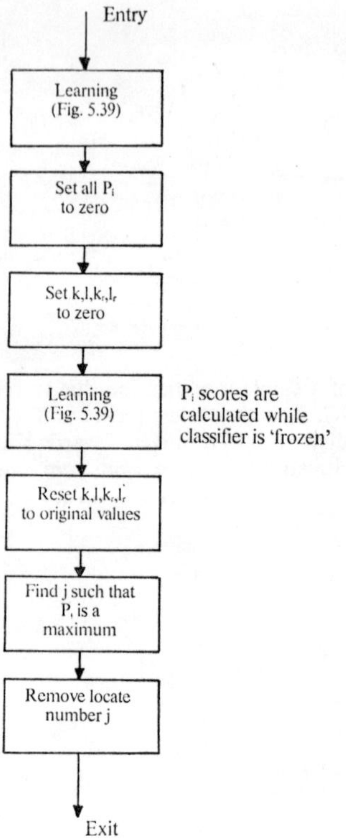

Figure 5.43 Procedure for pruning a multi-category NNC. Notation as in Figure 5.39.

of a CC. As before, the scoring processes are central to understanding these rules. From Figure 5.39 we see that:

(a) G_i records the frequency of modifying $\mathbf{B}_i$ during *error-correction* learning.
(b) P_i records the difference between the following quantities:
   (i) The frequency of modifying $\mathbf{B}_i$ when this is the closest locate, and the classifier decision is *incorrect*.
   (ii) The frequency of modifying $\mathbf{B}_i$ when this is the closest locate, and the classifier decision is *correct*.

5.13.3 Designing a Multi-Category PFC

Similar learning, growing and pruning rules have been developed for a multi-category PFC. The only major difference between these, and the procedures just described for a NNC, is in growing. In a NNC the decision surface is not significantly disturbed by the addition of a new locate, provided that it is placed close to an existing locate. In a PFC the potentials are additive, and so the decision surface could be seriously altered by the presence of the new locate. To avoid this, we should reduce the effect of both of these locates by halving their weights.

Other less significant changes are required because the weights in a PFC and NNC act in the opposite sense. In a NNC, the reduction of a weight increases the region of influence around a locate. In a PFC, this effect is reversed. To accommodate this difference, we must reverse the sign of the increments on the weights in both error correction and reinforcement learning.

5.13.4 Experimental Results

So far we have been unable* to test the rules outlined above, on anything other than type A data in two dimensions. The results of four experiments are summarised in Figures 5.44–5.46. Further notes are provided below:

(1) Table 5.1 shows the values of the parameters used in the experiments.
(2) Error rates were estimated by re-examining the training set. Of course, it would be preferable to evaluate error rates using previously unseen data, in a test set. If the classifier has few locates

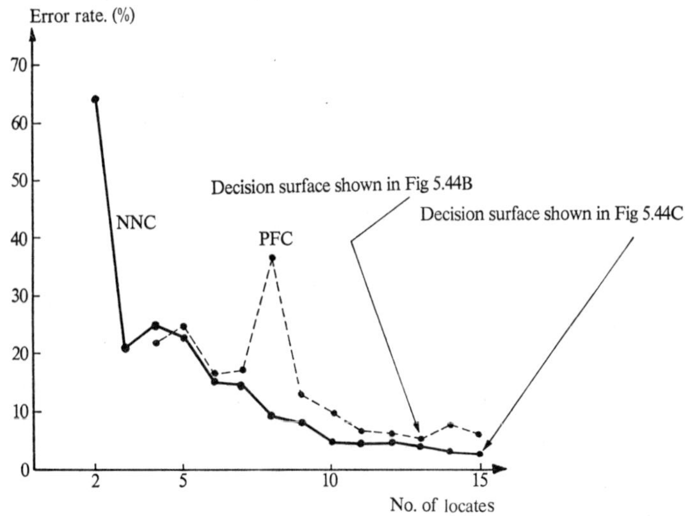

Figure 5.44 Results of growing a two-category NNC, and PFC using the procedure defined in Figure 5.42. Notice that the same teacher was used in these two experiments. (A) Error rates plotted against classifier size. (B) Decision surface of the teacher (solid), compared to that of the growing PFC with 13 locates (broken). Errors are indicated by stippling. The axes X_1, X_2 have been rescaled in the ratio 3:5. (C) Decision surface of the teacher (solid), compared to that of the growing NNC with 15 locates (broken). (Type A experiment. The details are given in Table 5.1. Source [13].)

* The only problem has been lack of time.

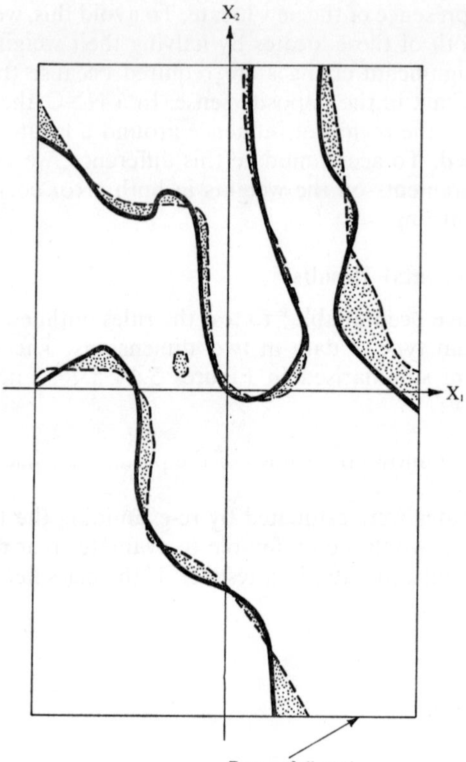

Figure 5.44 (*continued*)

compared to the number of points in the training set, the difference is likely to be small.

(3) Consider Figure 5.44; the values of the iteration parameters used in these experiments allow both reinforcement and error correction learning. Since k_r and l_r are relatively large, reinforcement learning predominates, when the error rate is less than about 20%.

(4) Notice the island of errors in the second quadrant of Figure 5.44B. There is one locate within this island, and it would be one of the first to be removed by pruning. Two small islands containing redundant locates can be seen in Figure 5.45B (bottom left-hand corner).

(5) Consider Figure 5.44A; the PFC initially had 4 locates. It is fortuitous that the initial positions of the locates were such, that they could all contribute usefully to the performance of the classifier. It is generally more satisfactory to start the growing

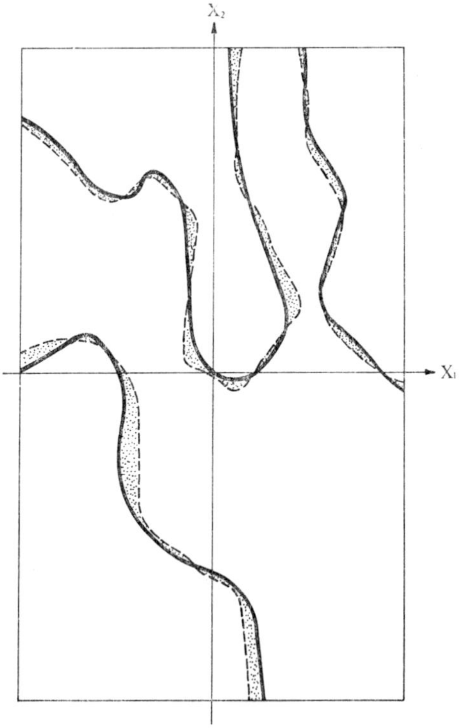

Figure 5.44 (*continued*)

process with one locate in each class. This reduces the likelihood of obtaining redundant locates.

(6) There was no reinforcement learning in the experiment summarised in Figure 5.45, since k_r and l_r were both zero.

(7) Consider Figure 5.45A; the error rate appears to have reached a limit of about 5–6%. This can be further reduced by adding reinforcement learning. The reason for this is unclear, and our research is continuing.

(8) Another important effect of reinforcement learning, is to force the locates further apart. This is an advantage in an analogue classifier, which is susceptible to noise.

(9) Consider Figure 5.46. k, l, k_r and l_r were all zero. Hence there was no reinforcement learning, and the weights were constant.

(10) Figure 5.46A should be interpreted as follows:

(a) the solid curve (sequence numbers 1–13) corresponds to growing,

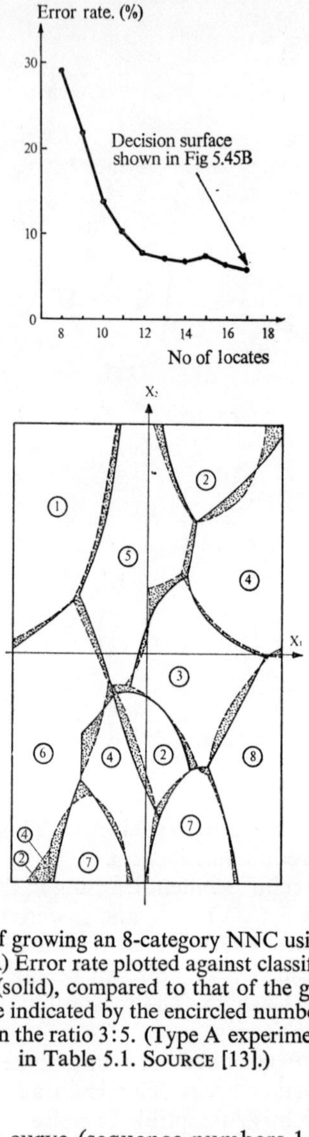

Figure 5.45 Results of growing an 8-category NNC using the procedure defined in Figure 5.42. (A) Error rate plotted against classifier size. (B) Decision surface of the teacher (solid), compared to that of the growing NNC with 17 locates. Class labels are indicated by the encircled numbers. The axes, X_1 and X_2 have been rescaled in the ratio 3:5. (Type A experiment. Details are given in Table 5.1. SOURCE [13].)

(b) the broken curve (sequence numbers 14–23) corresponds to pruning.

(11) Consider Figure 5.46A. Notice that a low error rate was first achieved with 12 locates (sequence number 10). However, pruning reduced the number of locates to 9, without increasing the error rate significantly. The "best" classifier, in terms of error rate/cost, is number 19.

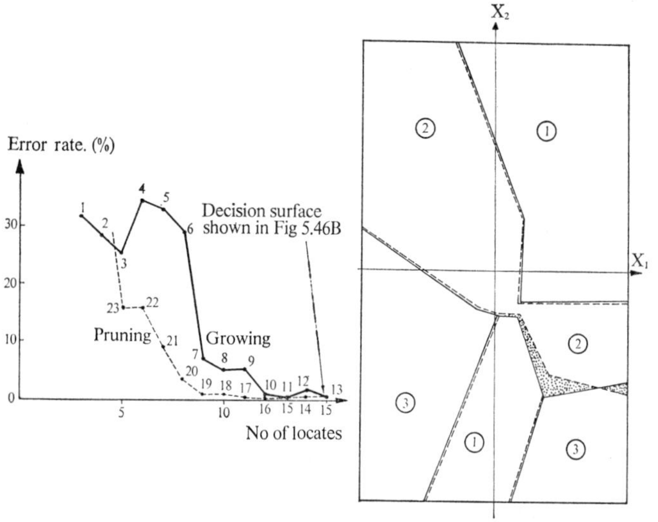

Figure 5.46 Results of growing and pruning a 3-category NNC, using the procedures defined in Figure 5.42 and 5.43. (A) Error rate plotted against classifier size. Numbers indicate the sequence of generating the various NNCs. The growing procedure was applied until there were 15 locates, after which pruning was applied. The "best" classifier, in terms of size and error rate, is number 19. (B) Decision surface of the teacher (solid), compared to that of the adaptive NNC with sequence number 13 (broken). Class labels are indicated by encircled numbers. The axes, X_1 and X_2, have been rescaled in the ratio 3:5.
(Type A experiment. Details are given in Table 5.1. SOURCE [13].)

Table 5.1 Details of the experiments whose results are shown in Figures 5.44–5.46. (SOURCE [13].)

	Figure 5.44B	Figure 5.44C	Figure 5.45	Figure 5.46
Type of Classifier	PFC	NNC	NNC	NNC
Number of Classes	2	2	8	3
Number of points from each class in the training set	950	950	500	900
Number of locates used in NNC simulating the teacher	16	16	15	7
Number of patterns examined during each learning period (MANY in Figure 5.39)	2500	2500	20 000	20 000
Initial number of locates in each class	2	1	1	1
k	0.05	0.05	0.05	0.05
l	0.001	0.001	0.01	0
k_r	0.01	0.01	0	0
l_r	0.001	0.001	0	0
Initial values of weights	1.0	1.0	1.0	1.0
Initial positions of locates	At the corners of the square surrounding the range of allowed patterns.			

REFERENCES

1. G. S. Sebestyen, Decision-making Processes in Pattern Recognition (Macmillan, New York, 1962).
2. N. J. Nilsson, Learning Machines (McGraw-Hill, New York, 1965).
3. A. G. Arkadev and E. M. Braverman, Teaching Computers to Recognise Patterns (Academic Press, London, 1967).
4. H. C. Andrews, Introduction to Mathematical Techniques in Pattern Recognition (Wiley, New York, 1972).
5. E. A. Patrick, Fundamentals of Pattern Recognition (Prentice-Hall, Englewood Cliffs, N.J., 1972).
6. T. M. Cover and P. E. Hart, Nearest Neighbor Pattern Classification, *Trans. I.E.E.E., IT-13*, 1, 21–27, 1967.
7. B. G. Batchelor, Learning Machines for Pattern Recognition, *Ph.D. Thesis*, Southampton, 1969.
8. W. K. Taylor, Pattern Recognising Adaptive Controller, *Proc. I.F.A.C. Conf. on Automatic and Remote Control*, 488–496, 1963.
9. B. R. Wilkins and B. G. Batchelor, Evolution of a Descriptor Set for Pattern Recognition, *Proc. Conf. on Technical and Biological Problems of Control, Instrument Soc. of America*, 794–9, 1970.
10. A. N. Mucciardi and E. E. Gose, Evolutionary Pattern Recognition in Incomplete Non-linear Multi-threshold Networks, *Trans. I.E.E.E., EC-15*, 2, 257–262, 1966.
11. B. G. Batchelor, Error Correcting Algorithm for Adapting a Hyperelliptical Decision Surface for Pattern Recognition, *Electronics Letters*, **4**, 10, 204–6, 1968.
12. B. G. Batchelor and B. R. Wilkins, Adaptive Discriminant Functions, *Pattern Recognition, I.E.E. Conf. Pub.*, **42**, 168–178, 1968.
13. B. G. Batchelor, Automatic Procedures for Varying the Complexity of a Pattern Classifier, in Machine Perception of Patterns and Pictures, *Inst. of Physics, Conference Series*, No. 13, 317–327, 1972.
14. B. G. Batchelor, Growing and Pruning a Pattern Classifier, *Information Sciences*, **6**, 97–122, 1973.
15. B. G. Batchelor and B. R. Wilkins, Method for Location of Clusters of Patterns to Initialise a Learning Machine, *Electronics Letters*, **5**, 20, 481–3, 1969.
16. N. L. Ford, B. G. Batchelor and B. R. Wilkins, Learning Scheme for the Nearest Neighbour Classifier, *Information Sciences*, **2**, 139–157, 1970.
17. B. G. Batchelor, N. L. Ford and B. R. Wilkins, Learning in a Potential Function Classifier, *Electronics Letters*, **6**, 25, 826–8, 1970.
18. B. G. Batchelor, Hybrid Pattern Classifier for Computer Input, in Computer Science and Technology, *I.E.E. Conf. Pub.*, No. 55, 141–150, 1969.

Chapter 6

Practical Details and Implementation of Classifiers

6.1 INTRODUCTION

So far in this book, we have discussed the classification and learning schemes in isolation from the real world. There is a large gap between our present theoretical understanding of the techniques, and the ability to apply them successfully. There are many minor details, which when taken together, can greatly influence the results of a classifier design programme. We list these minutiae in this chapter, and also discuss the implementation of the classification and learning schemes.

6.2 PRACTICAL ASPECTS OF CLASSIFIER DESIGN

In this section we list those details which are necessary for designing a CC.* (Also see Appendix 1.)

6.2.1 Initial Number of Locates

Start growing with one locate. If more than one is used, we often find that most of them degenerate. If a really effective cluster analysis program were available, we could remove this constraint.

6.2.2 Initial Position of Locates

Assuming that we do not possess a really effective cluster analysis procedure, the single locate is best placed at a point typical of the class characterised by $T = +1$. Such a point may be found by selecting the mean vector of this set. Although crude, this technique is quite effective.

6.2.3 Initial Value of Size Parameter

The initial choice of size parameter should be such that the hypersphere (or its equivalent for non-Euclidean distances) is small. Both this, and the choice of initial locate position, described above, ensure that the hypersphere does not enclose large voids (Figure 6.1).

* The author is more familiar with this classifier than any other, and therefore feels that he is better able to do justice to this important subject, if he restricts the discussion to the CC. Other factors have influenced this decision; the difficulty of implementing the ϕ-machine, and instability in the NNC/PFC.

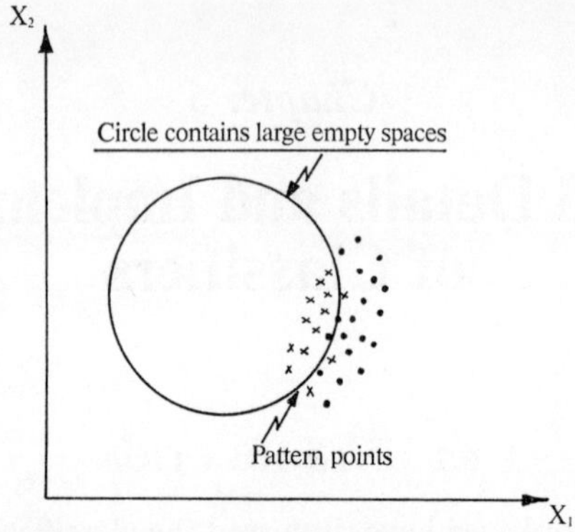

Figure 6.1 A situation that may arise if the initial choice of locate position and size parameter were unsuitable. If we add another circle by fission, and then attempt to optimise this enlarged CC by learning, we should obtain very little improvement in the error rate. On the other hand, if we perform this operation starting with a smaller circle, we should obtain better results.

6.2.4 Maximum Number of Locates

While we are growing a classifier, we increase its size (number of locates). In principle, there is no reason why we should not continue this process indefinitely. However, we shall eventually find that we can separate the training set perfectly, but when we attempt to apply our classifier to new data, we shall obtain poor results. In other words, any 2 *finite* sets can be separated if the classifier contains enough locates. A parallel situation exists in fitting a polynomial curve to graphical data. If there are, say, n points on a 2-dimensional graph, then a polynomial curve of the order $(n - 1)$ can be made to intersect all of the points.

An approximate analysis [1] has shown that any two sets, each containing h points, may usually* be separated by a CC which has N locates, where:

$$N \geq \frac{h}{q + 1} \tag{6.1}$$

(As usual q is the dimensionality of the **X**-space.)

These sets may usually be separated, even if they were generated at random with the same probability distributions. If the CC contains fewer than $(h/(q + 1))$ locates it is unlikely that a randomly generated training set can be successfully separated. To avoid the danger of producing an unduly large and inefficient classifier, we should keep the number of locates in our CC below $(H/2(q + 1))$, where H is the total population of the training set. This assumes that the training set contains $\frac{1}{2}$H patterns from both of the classes. If, as often happens, the

* With a probability greater than 0.5.

training set contains unequal numbers of patterns from the two classes, then we should not use more than $(h'/(q + 1))$ locates, where h' is the population of the class represented less frequently within the training set. This is the maximum value, and it is advisable to restrict the size of the classifier to be much smaller than this. We have found no difficulty, if the CC has fewer than $(h'/10(q + 1))$ locates.

6.2.5 Size of the Data Sets

Unfortunately, it is not possible to state precisely how large the data sets should be, *before* we begin our analysis. However, we can decide whether the training and test sets were large enough, *after* the analysis is complete. The criteria we apply to make this decision are listed below:

(a) *For the test set:* Is the test large enough to enable us to estimate the error rate of the classifier, by counting the number of disagreements between it and the teacher? (see Figure 6.2)

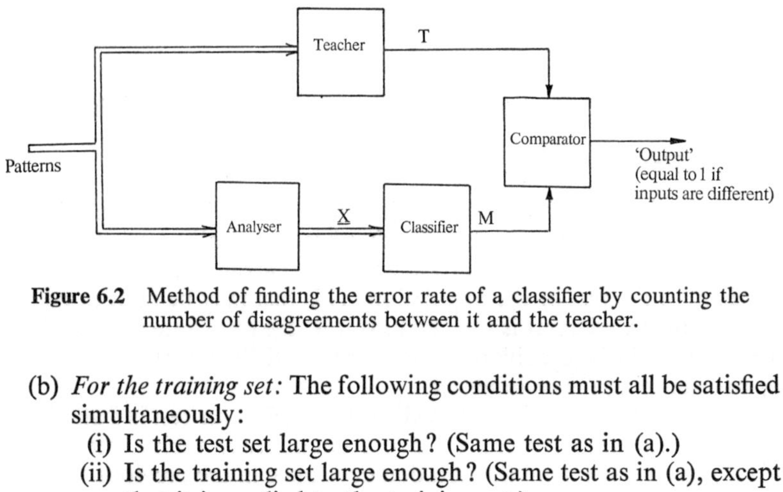

Figure 6.2 Method of finding the error rate of a classifier by counting the number of disagreements between it and the teacher.

(b) *For the training set:* The following conditions must all be satisfied simultaneously:
   (i) Is the test set large enough? (Same test as in (a).)
   (ii) Is the training set large enough? (Same test as in (a), except that it is applied to the training set.)
   (iii) Is the error rate measured with the training set, approximately the same as the error rate, measured with the test set?

6.2.5.1 Size of the Test Set

We must test each class separately. This will avoid the situation where the test set contains an adequate sample of one class, but not the other. Let us denote the number of representatives of a given class in the test set by P_1. When these patterns are presented to the classifier and teacher, there will be (say) D_1 disagreements. That is, on D_1 occasions the classifier and teacher produce different decisions, and on $(P_1 - D_1)$ patterns they agree. The ratio (D_1/P_1) is taken as the estimate of the error rate, for the single class being considered. Conventional statistical techniques may be used to find the *confidence interval* for this estimate (*e.g.* [2], page 99 and table 4). This interval represents that range of values of the *true* error rate (measured on the complete population), which could reasonably be expected to yield D_1 disagreements out of P_1 trials. In order to make an accurate estimate of the true error rate, the confidence interval should be small compared to the estimated value, (D_1/P_1). Hence the width of the confidence interval should not be greater than (about) 10% of the ratio D_1/P_1. If it is, then P_1 is too small and must be increased. This test is repeated for both pattern classes.

Although the technique described above has the benefit of mathematical respectability, a far simpler rule of thumb can be derived. This simply states that P_1 is sufficiently large if the number of disagreements, D_1, exceeds 100. This guarantees that the *standard error* of the estimate for the error rate is less than 10%. Hence, in order to obtain a reasonably accurate estimate of an error rate, which is equal to $100.\lambda\%$ (probability $= \lambda$) we need

$$\frac{100}{\lambda} \tag{6.2}$$

patterns. Do not forget that this test must be applied to both classes.

6.2.5.2 *Size of the Training Set*

If the training set is very small, it will not be truly representative of the parent populations. A classifier that has been trained using a small training set will therefore yield a much smaller error rate on the training set than on the rest of the population. This fact may be used to detect whether the training set is too small. If it is, the error rate measured on the test set will be considerably greater than the error rate established by re-examining the training set.

Again, we must consider each class separately. Our test is a standard statistical technique, based on 2 × 2 contingency tables (see [2], p. 141–3). We must extend the notation developed in the previous section. Let:

P_1 and D_1 be as in the previous section (they refer to the test set);
P_2 be the number of patterns from the given class in the *training* set;
D_2 be the number of disagreements from the given class in the *training* set.

In order for our test to be valid, we must ensure that each of the following quantities is greater than 5:

$$D_1,\ D_2,\ (P_1 - D_1),\ (P_2 - D_2)$$

We then compute the following quantity, which we shall test against a chi-square distribution:

$$\frac{|(P_1 + P_2)(P_1D_2 - P_2D_1)|}{P_1P_2(D_1 + D_2)(P_1 + P_2 - D_1 - D_2)} \tag{6.3}$$

The following percentiles of the chi-square distribution, with one degree of freedom are taken from standard tables:

percentiles	*confidence coefficient* (%)
0.00016	1.0
0.0039	5.0
0.0158	10.0
1.323	75.0
2.71	90.0

Assuming that our test of validity has been satisfied, we compare the value of expression (6.3) to the list above. This will indicate whether there is a significant difference between the error rates measured using the test and training sets. Of course, an acceptable level of confidence is difficult to define; the choice depends upon the individual's judgement.

The above test must be applied to both classes separately, and both classes must yield acceptable (that is small) values for (6.3).

The author feels that it is worthwhile exercising caution, by repeating the test described in section 6.2.5.1, on the training set. This simple test requires that D_2 is greater than 100.

6.2.6 Order of Presentation of Data

It is important to ensure that data is presented to the classifier during learning in an unstructured (random) sequence. We might be tempted, for ease of programming, to present all members of one class to the classifier, and then present the other class afterwards. This is unwise, as it will lead to oscillation (Figure 6.3). It is quite acceptable to present members of the two sets alternately.

For the same reason, it is desirable to shuffle the data prior to learning, or to use a random data selector during learning. This will avoid the possibility of oscillation, due to correlations between patterns. It is essential to shuffle the data when it arises from a time series, for example, an EEG recording. It is also important when there is no obvious structure, to avoid the effects of unknown (or forgotten) associations between patterns. A practical selection procedure is shown in Figure 6.4.

6.2.7 Deciding Which Class Should be Enclosed Within the Hyperspheres

Consider Figures 5.33A and B. It would be sensible to enclose the more compact class (Figure 5.33A) within the circles of a CC. Two circles would be required, compared to the three circles, which would be needed, if the class shown in Figure 5.33B were enclosed.

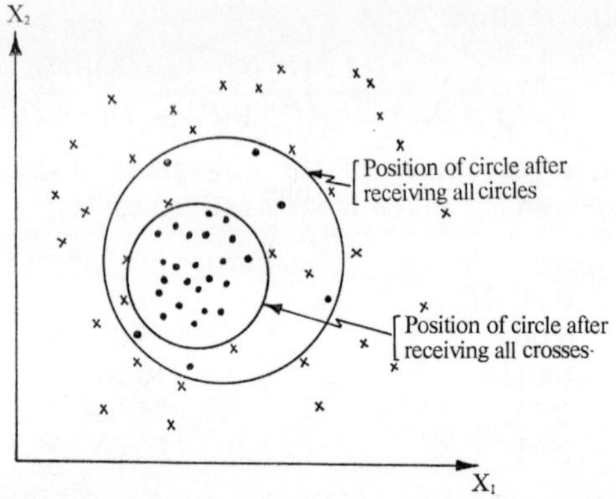

Figure 6.3 Oscillation of a decision surface due to a non-random order of data presentation. Suppose all patterns in one class are presented to the classifier, before presenting all patterns from the other class. After an initial period when the classifier is learning rapidly, the decision surface will oscillate.

The problem of deciding which is the more compact class is quite difficult in high-dimensional spaces. It is conceivable that cluster analysis techniques could be used, although it would probably be simpler to apply the growing/pruning procedure twice, with the class labels reversed.

6.2.8 Choice of Learning Rules

In section 5.6 (equation (5.6)) we suggested a locate modification rule which we claimed to be simpler to implement than the original (Box A in Figure 5.10). This is almost certainly so, provided we are considering a hardware implementation of the learning rules. In software, the reverse is probably true.

We have found very little difference between the results obtained using these rules, and suggest therefore that we use whichever is the simpler to implement.

6.2.9 Choice of Distance Measure

The only reason for ever contemplating the use of distances other than Euclidean, was that this was expensive to implement in electronic hardware. The advent of both analogue and digital integrated circuits has enabled us to review the situation. It is only since 1970, that digital or analogue squaring circuits have been available, at a price that is feasible for their inclusion in a pattern classifier. Nevertheless, squaring is still a difficult operation compared to the functions inherent in calculating the square or city block distances.

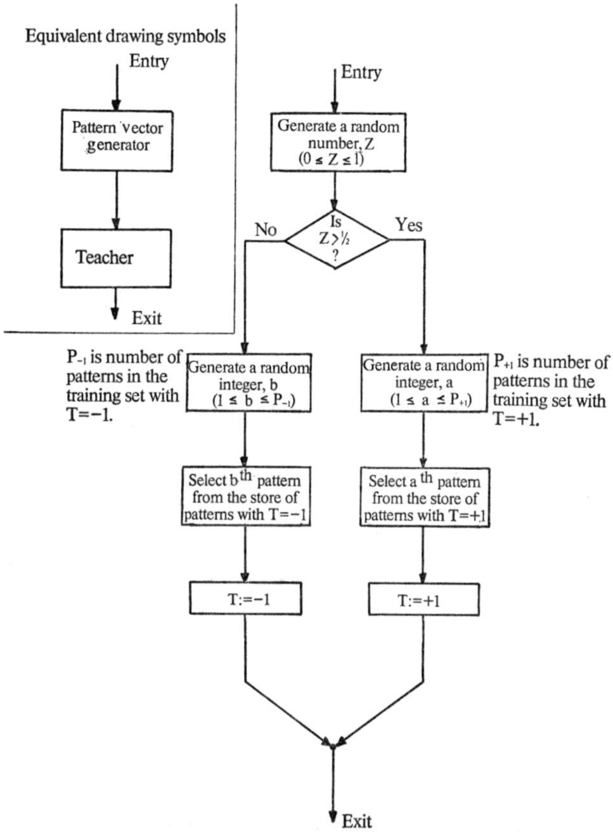

Figure 6.4 A random data selector. All three random number generators are independent and produce uniformly distributed outputs.

There are certainly good reasons for avoiding the city block and square distances in a NNC or PFC, but until now we have tacitly assumed that they could be employed in a CC. Problems can arise during fission as we illustrate in Figure 6.5, with reference to the city block distance. The solid square was fitted by learning, while the broken one was added during fission. Since all of the pattern points are contained within the dotted curve, the only parts of the squares that are being used are the two upper, right-hand sides. Since these sides are parallel, there has been no increase in the effective complexity of the decision surface, following fission. This situation cannot be resolved by growing, as more squares will be added on top of these two. The situation is completely avoided in the CC using the Euclidean distance, since the hypersphere has no straight sides.

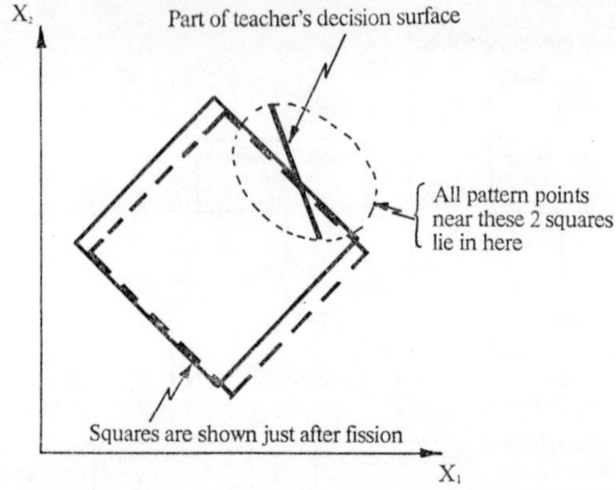

Figure 6.5 Emphasising the disadvantage of the non-Euclidean distances in the CC during fission. This situation occurs just after fission. There has been no effective increase in the complexity of the decision surface. Both squares present linear decision surfaces within the populated region (enclosed by the dotted curve) and these are parallel. Hence, there is only one effective linear surface, which cannot be rotated (except by gross movements of the locates to introduce a corner into the populated area). There may be other parts of the space which contain pattern points, but these are assumed to be a long way from the region shown here.

6.2.10 Multi-Category Pattern Classification Using a CC

We have so far ignored the question of providing a multi-category classification using CCs, although the solution is quite straightforward. Suppose we wish to distinguish between the classes A, B, C. We should first of all use a CC to learn to distinguish between class A on one hand, and the set consisting of the union of classes B and C, on the other. We should then train another CC to separate B and (A ∪ C). Finally, we should train a CC to discriminate between class C and (A ∪ B). We then have three classifiers which can be operated in parallel (Figures 6.6 and 6.7). The possibility now exists of obtaining simultaneous outputs from two or more of the classifiers, indicating that the pattern could lie in more than one class. Such a state of confusion can be recognised using a linear classifier (Figure 6.6).

6.2.11 Minor Modification to Fission

In Figure 5.31, we indicated that during fission, the new locate $\mathbf{B}_N$ would be added close to the locate $\mathbf{B}_i$, using the rule:

$$\mathbf{B}_N = \{B_{i,j} + E\} \tag{6.4}$$

where

$$\mathbf{B}_i = (B_{i,1}, B_{i,2}, \ldots, B_{i,q}) \tag{6.5}$$

The quantity E was stated earlier to be a small number which we implied (by default) was constant. It would be better to vary E, to avoid

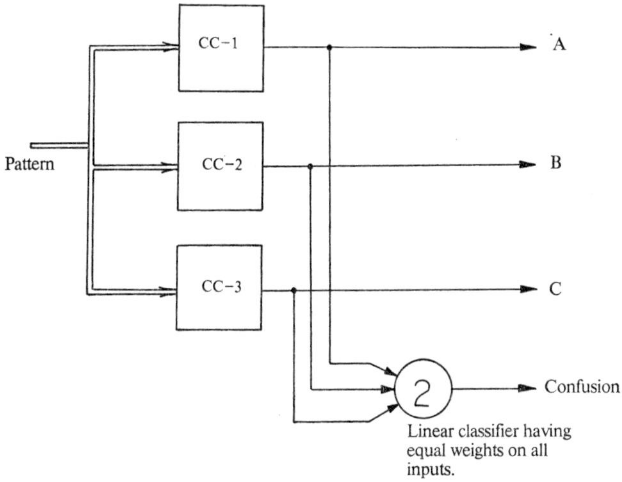

Figure 6.6 Construction of a 3-category classifier which has a "don't know" (confusion) output. CC-1 to 3 are normal 2-category CCs.

the continued enlargement of the classifier, without an improved performance. A situation in which this would be an advantage is illustrated in Figure 6.8A. We assume that, by some means, we have obtained a classifier consisting of the solid circle and other circles, which partially overlap it (cross-hatched area). Suppose, we now find that the solid circle should be subjected to fission and that the new circle (broken curve) is placed as shown. Patterns outside both the solid and broken circles will now cause only the solid circle to enlarge (point R

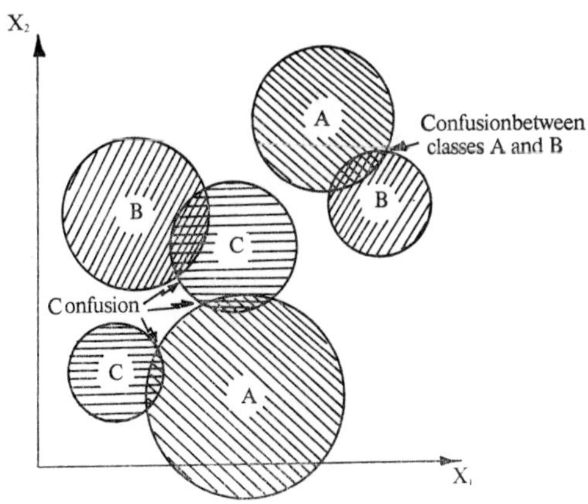

Figure 6.7 Decision surface of a multi-category classifier of the type outlined in Figure 6.6.

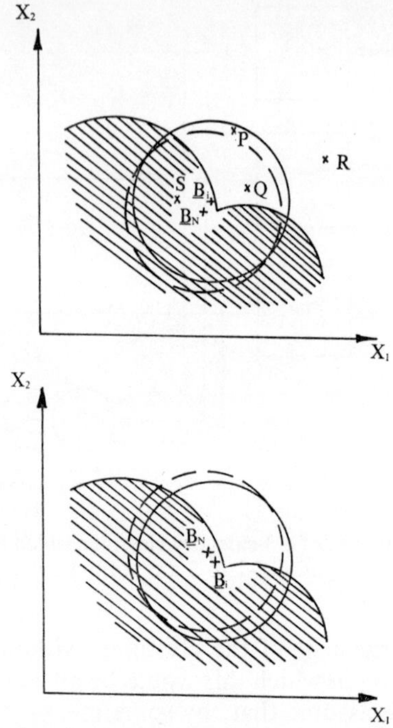

Figure 6.8 Illustrating the need for randomly chosen positions for the new locate during fission. (A) The broken circle will become smaller and will never contribute usefully to the decision surface. (B) A more satisfactory situation. Both of the non-shaded circles can now be modified during learning and will contribute to the decision surface in a useful way.

is typical). Points such as S and Q will lead to the reduction in the diameters of both circles, while points typified by *P* will reduce the diameter of only the solid circle. The broken circle will become smaller, since there are no points which cause it to increase its diameter. In contrast, the solid circle remains in approximately the same equilibrium position as it was, prior to fission. The addition of the new circle has not improved the power of the classifier. If the new circle had been added at the position indicated in Figure 6.8B, this difficulty would have been avoided. Normally, it would be difficult to determine the best position for the new locate. We have used a random number generator to select the orientation of $(\mathbf{B}_N - \mathbf{B}_i)$ (Figure 6.9). This solution appears to be quite satisfactory.

6.2.12 Choice of Iteration Parameters

At the end of a long learning sequence, the classifier is in a state of dynamic equilibrium, with a small tremor on its locates and weights. The magnitude of this vibration increases with *k* and *l*. When we stop learning, we fix the classifier at some arbitrary point in this vibrating

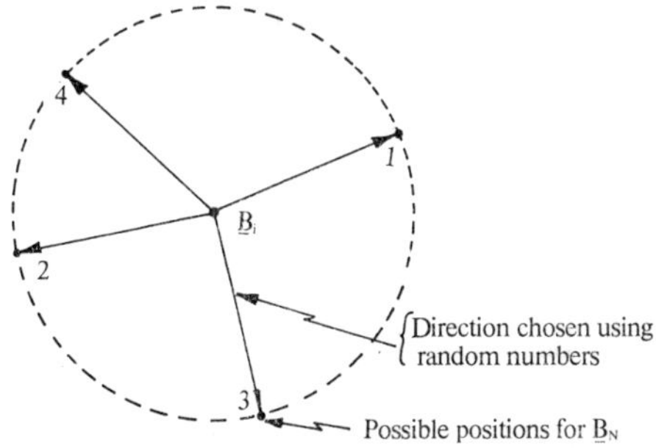

Figure 6.9 Showing random generation of new locates during fission. $\mathbf{B}_i$ is that locate which had the highest growing score, prior to fission. A new locate $\mathbf{B}_N$ is to be added close to $\mathbf{B}_i$. If the first additional locate (1) is unable to increase the effective capacity of the CC, then a different direction is chosen to generate the next new locate (2).

equilibrium state. The accuracy of this fixed classifier is likely to be reduced if k and l are large. On the other hand the learning will take an unduly long time if very small values of k and l are used.

In section 5.6 we indicated that the following range of values is suitable:

$$0.0001 \leq k \leq 0.05$$

$$0.001\lambda \leq l \leq 0.05\lambda$$

where λ is the diagonal of the smallest window containing the pattern points.

Certainly, we should never use larger values, although if computer time is available, smaller values can be used. It should be possible to reduce k and l progressively, in such a way that we obtain rapid learning, with a precise classifier available afterwards. We do not yet know how this should be done to achieve optimal results.

6.2.13 Number of Iterations During Learning

With the iteration parameters within the range specified above, we have never encountered any significant changes occurring after about 5000 iterations (Figures 5.23 and 5.24). For safety we should allow more, say 10 000 iterations, which will ensure that learning is complete, before we add or remove locates.

6.2.14 Deciding Whether to Grow or Prune

We should begin by growing a CC until one of the following criteria is satisfied:

(i) There is no significant improvement in the performance as the classifier size is increased (human judgement is quite adequate).

(ii) The error rate reaches the best obtainable results, predicted by the technique described in section 5.4.1.
(iii) the classifier is too large, compared to the quantities of data available (see section 6.2.4).

When any of these conditions is satisfied, we revert to pruning. Often during pruning, we shall observe a fairly sharp "elbow" in the graph of error rate plotted against classifier size (Figures 5.37 and 5.38). This point indicates the "best" classifier. When there is no such elbow, some other criterion for the "best" classifier must be used. This may be specified by the problem, but in the absence of any other criterion, we could use the product of classifier size and its error rate to assess its merit.

Overgrowing (and overpruning) is never serious as the situation can always be overcome by pruning (or growing).

6.3 IMPLEMENTATION OF PATTERN CLASSIFIERS

We have, until now, regarded a pattern classifier as if it were a spiritual being with no physical existence. Pattern recognition will never be successful unless we can build *machines* that are very fast and cheap. Nevertheless, where there is a plentiful supply of computer time, it may be preferable to use a software realisation, rather than hardware.

Our approach was originally directed towards analogue hardware, although the recent, rapid development of digital integrated circuits has offered scope for new forms of implementation. We have a three-way choice:

(i) software,
(ii) analogue hardware,
(iii) digital hardware.

Of course, each has its own particular merits so we must describe the implementation of a pattern classifier using all three methods, leaving the reader to make the final selection.

6.4 SOFTWARE IMPLEMENTATION

Appendix 1 presents a flow-chart for designing a compound classifier.* The growing and pruning procedures are specified at a sufficiently low level that they can be coded directly into Algol or Fortran.

The reader is warned that writing software to implement a learning classifier is a very difficult task. The reason is that the self-correcting

* We have tried to collect together the hints discussed earlier in the chapter, so that the reader can design a classifier with as little human intervention as possible.

properties of the program often obscure errors in the programming or data. This problem cannot be overstated. Self-adaptive programs are always difficult to debug, so this task should always be approached with great care. It is made somewhat easier by the use of an interactive computing system. Errors in the learning program will reduce the accuracy of the final classifier, but its error rate can be measured quite reliably using a separate program to simulate the final, fixed classifier. This is important, since it allows checks to be made independently of the learning program.

6.5 ANALOGUE IMPLEMENTATION

For the moment, we shall restrict our discussion to the implementation of a CC using analogue hardware, without reference to the question of realising the learning procedures. We shall see in section 6.8 that this can be accomplished using a hybrid system, incorporating a hardware classifier, and a general-purpose digital computer.

The rapid advance of integrated circuit techniques has led to a significant change in our approach to the problem of implementing pattern classifiers. Until recently the calculation of the Euclidean distance would have been prohibitively expensive. We can now buy an analogue multiplier for under five pounds, which is cheap enough to allow us to forsake all distance measures other than Euclidean. We have seen throughout this book, that non-Euclidean distances, although widely supported in the literature, suffer from a number of disadvantages. So we are prepared to advocate the use of the Euclidean distance, realising that the cost of implementing it is unlikely to rise.

Figure 6.10 shows the circuit diagram of an analogue multiplier, based on the SG1495 device, and which costs about £2.50 (July 1972). The characteristics of this multiplier are summarised in Figure 6.11. Figure 6.12 shows the method of implementing the complete hypersphere subclassifier, while Figure 6.13 shows how its decision surface (in 2 dimensions) may be drawn automatically. Figure 6.14 shows the circular decision surface obtained in this way, using the subclassifier based on the SG1495.

The "frequency response" of the classifier (see Figure 6.11) is limited by the low grade operational amplifiers. The squaring operation is accurate up to 10 kHz (see Figure 6.11). The minimum response time of the classifier is therefore approximately 100 μs. The response time of a parallel system is, of course, independent of the number of dimensions, and the number of subclassifiers. The cost is approximately £5/dimension/subclassifier.

In a fixed CC the locates and size parameters may both be set by resistor chains (Figure 6.15). This is cheap, stable and allows a classifier to be used in a multitude of tasks by simply exchanging component cards.

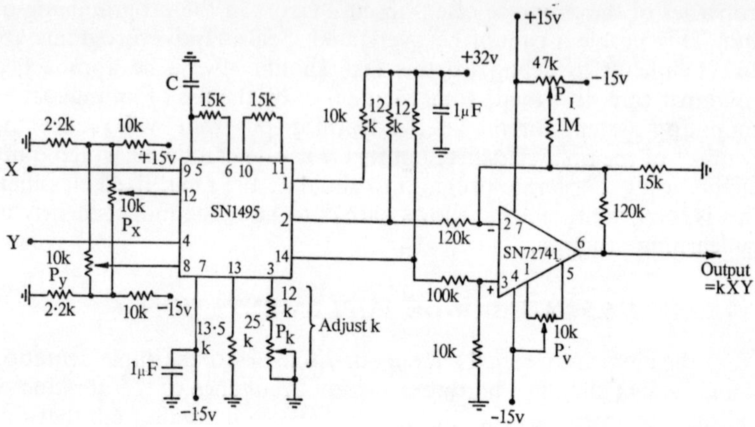

Figure 6.10 Circuit diagram of an analogue multiplier. The inputs are X and Y.

Input voltage levels: $-10\text{ V} \leq X, Y \leq +10\text{ V}$
Output voltage level: $-10\text{ V} \leq KXY \leq +10\text{ V}$
Maximum input frequency about 10 kHz.
Controls:

P_I, current offset;
P_V, voltage offset;
P_X, X offset, adjust;
P_Y, Y offset, adjust;
P_K, K adjust;
C must be set for optimum frequency response.

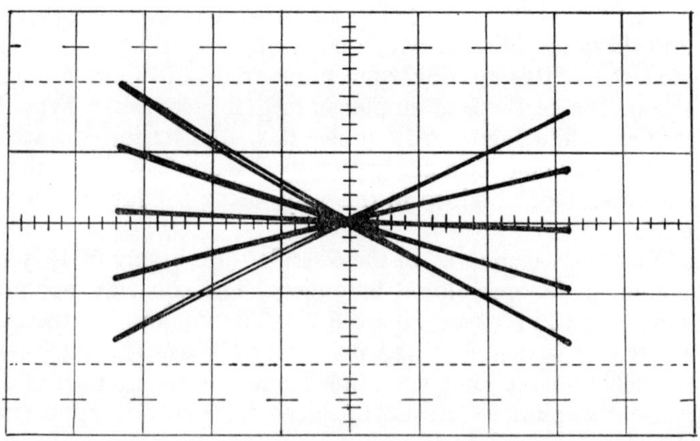

Figure 6.11 Characteristics of the analogue multiplier. (A) Superimposed input-output characteristics, with one input preset to the values 0 V, ±5 V and ±10 V. Sensitivity 2 V/division, measured at 2.5 kHz. (B) Wired as a squaring circuit. Sensitivity 2 V/division. Measured at 10 kHz. (C) Wired as a squaring circuit. Sensitivity 2 V/division. Measured at 25 kHz.

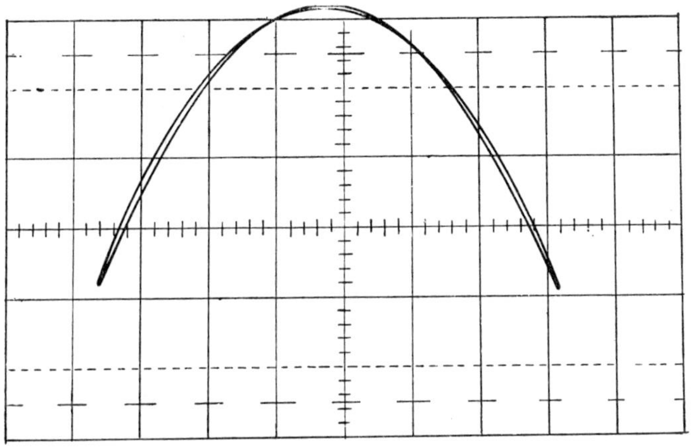

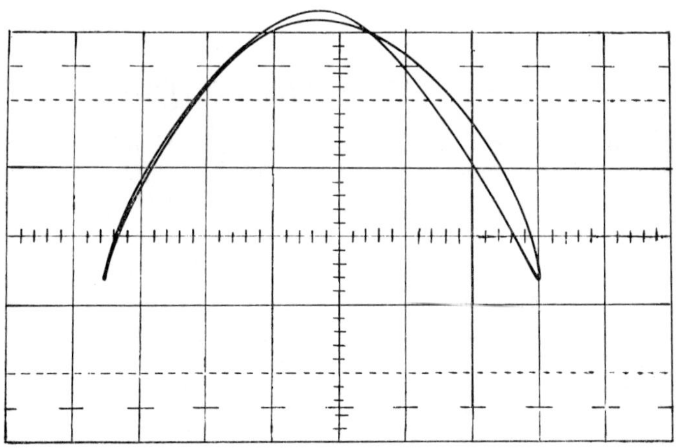

Figure 6.11 (*continued*)

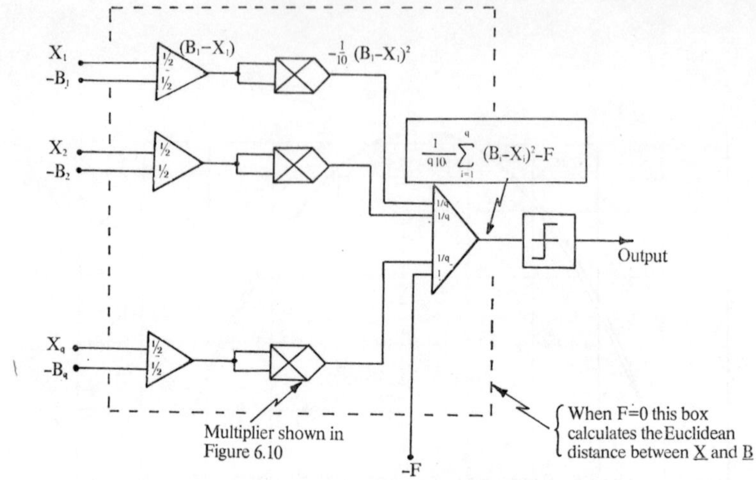

Figure 6.12 Analogue realisation of a hypersphere classifier. X_i and B_i may vary in the range -10 V to $+10$ V.

The implementation of the modified NNC requires fairly straightforward computing techniques in addition to the calculation of the Euclidean distance. A system is shown in Figure 6.16.

6.6 DIGITAL IMPLEMENTATION

Until the advent of cheap integrated circuits, the implementation of a CC using digital techniques would have been prohibitively expensive. It is now possible to build a digital version of the CC that is little more expensive than the analogue circuit described above. The greater cost

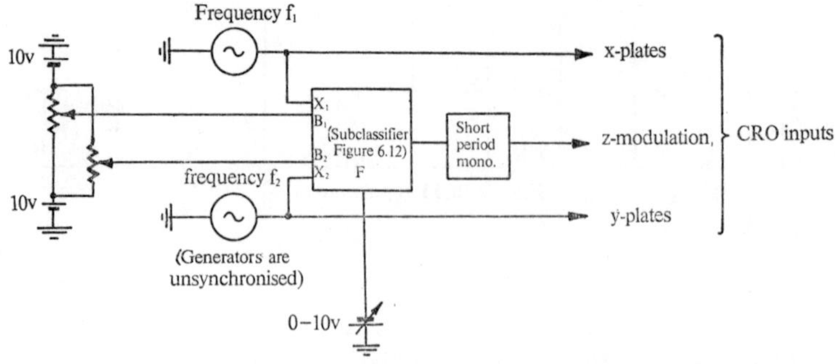

Figure 6.13 Method of drawing the decision surface of a classifier on a CRO.

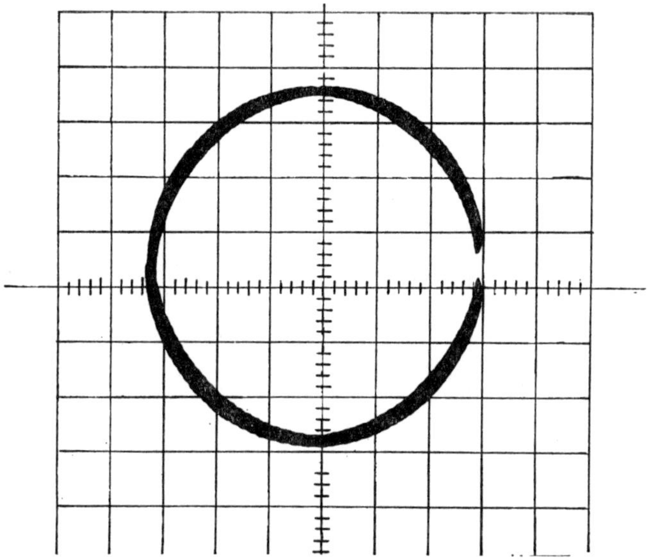

Figure 6.14 Decision surface of a circle subclassifier, drawn by the technique outlined in Figure 6.13. Parameters relate to that figure: $f_1 = 26$ Hz, $f_2 = 26.3$ Hz, $F = 3.3$ V, $-6\text{ V} < X_1 < +6\text{ V}$, $-6.4\text{ V} < X_2 < +6.4\text{ V}$.

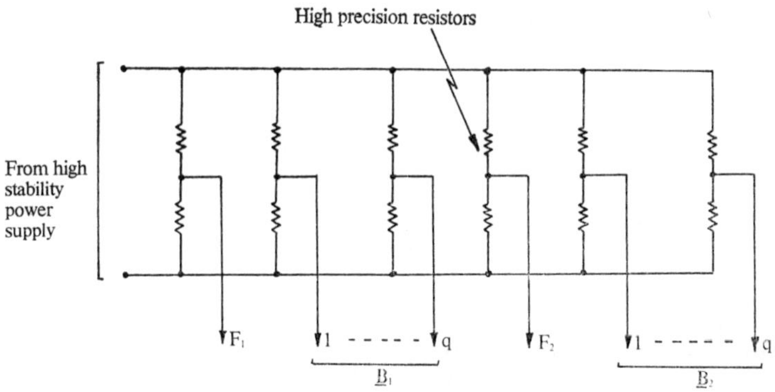

Figure 6.15 Method of storing locates and size parameters in an analogue classifier.

of the digital implementation is unimportant, when one realises that greater speeds can be achieved with reduced drift.*

* A digital system is, of course, free from drift. However in many applications the descriptors would be derived from an analyser which provides analogue electrical signals. In this case, it would be necessary to use an analogue-digital converter on the classifier input. This will introduce some drift.

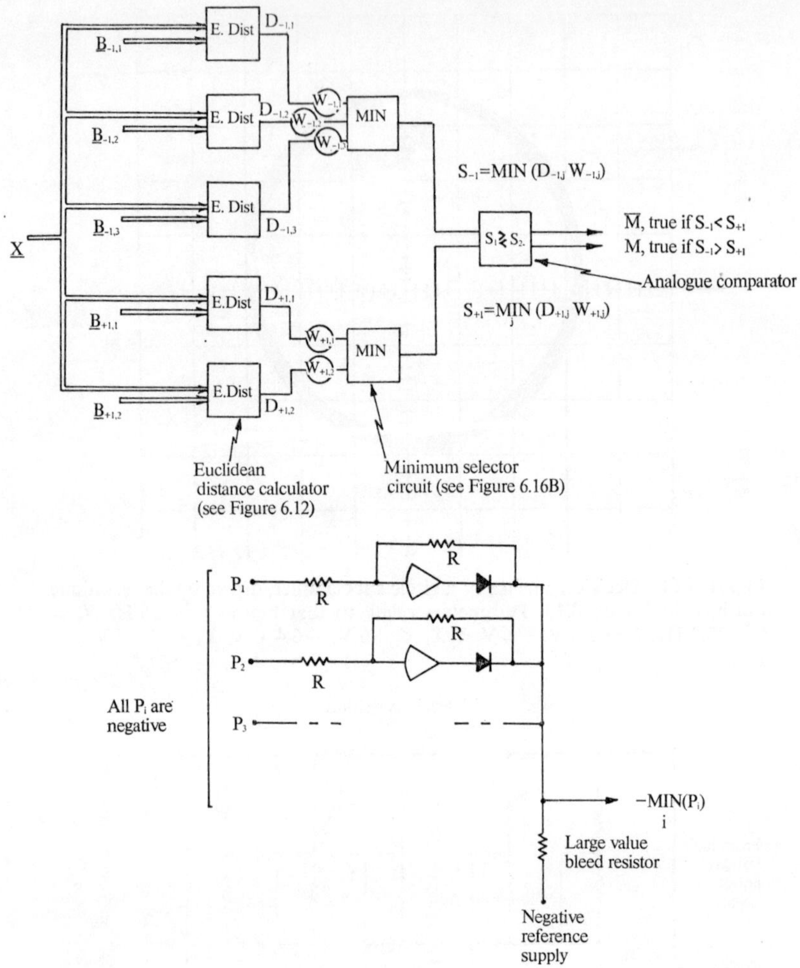

Figure 6.16 Method of implementing a modified NNC. (A) Overall system. (B) Minimum selector circuit to find min (P_i), where P_i corresponds to the weighted distances in (A).

In October 1971 one of the author's students, P. G. Ellis,† began a study of the digital implementation of a CC. He designed and built a simple classifier, similar to that shown in Figure 6.17. Some of the details differ from his original circuit. Modifications have been made to allow easier explanation and to accommodate more dimensions in the **X**-vector. Table 6.1 lists the operations required by the circuit shown in Figure 6.17, to realise a hypersphere classifier. Table 6.2

† The author acknowledges with gratitude the contribution made by Peter Ellis who conducted this work for a final year undergraduate project.

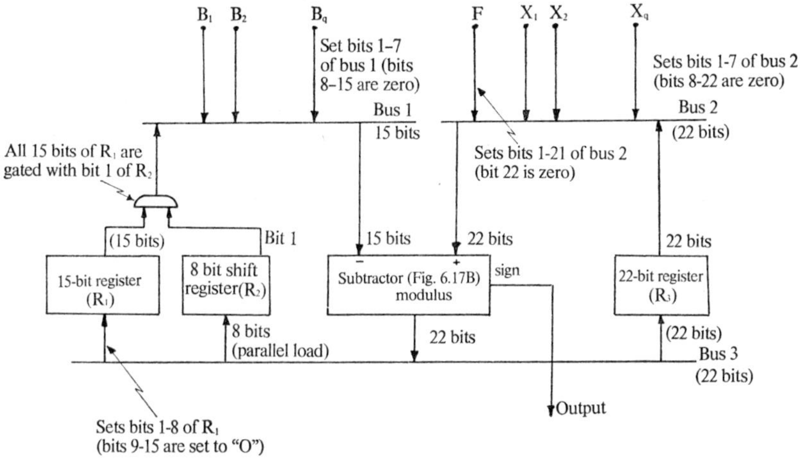

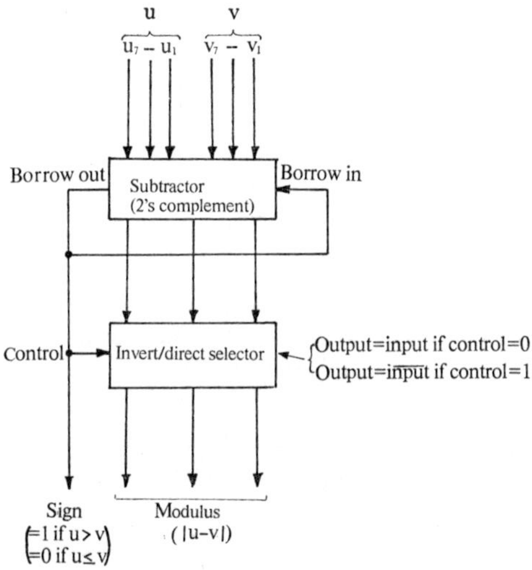

Figure 6.17 Serial-parallel digital implementation of a hypersphere subclassifier. The operation is explained in Table 6.1. The number of dimensions, q, must not exceed 32. (A) Complete system. (B) Detail of the subtractor, required in (A).

summarises the performance of the original circuit. By July 1972, the cost of digital multipliers and adders was such that a fully parallel arrangement was feasible (Figure 6.18, Table 6.3). Although the fully parallel system is more expensive, it is very much faster than the serial-parallel circuit.

The locates and weights must be stored by the classifier. In a digital implementation, we must resort to quite complex storage techniques,

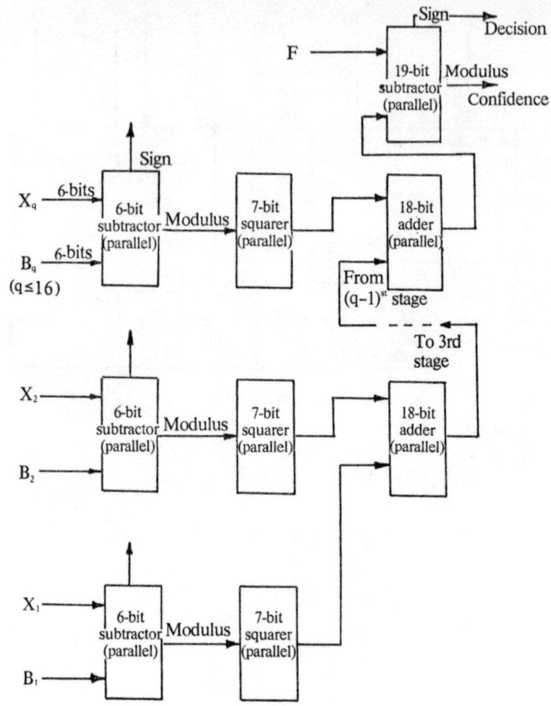

Figure 6.18 Fully parallel digital implementation of a hypersphere sub-classifier. The subtractors have a similar structure to that shown in the preceding figure.

compared to the fixed resistors employed by an analogue classifier. The storage must be permanent to avoid accidental erasure during power failures. A programmable read only memory is most convenient.

In many applications information will be presented to the classifier in analogue form. In these situations we must employ analogue-digital converter(s), if we are to use a digital classifier. The cost of this conversion greatly increases the overall cost of the system. Of course, there are many applications in which input information arrives in a form unsuitable for an analogue classifier (for example, on paper tape). Again in these situations, we must add the cost of information recoding, before we can compare costs properly.

6.7 IMPLEMENTATION USING MICROCOMPUTERS

The present generation of microcomputers is perfectly adequate for most of our purposes. We could, for example, use a microcomputer to implement the growing and pruning rules for a CC. In many situations,

Table 6.1 Showing operations required by the serial-parallel digital system for implementing a hypersphere classifier (Figure 6.17).

Step	Operation	Comments
1	Reset	Set all registers to zero.
2	$R_3 := \mathrm{F}$	Store F in R_3.
3	$R_1 := \lvert X_1 - B_1\rvert$ $R_2 := \lvert X_1 - B_1\rvert$	Store $\lvert X_1 - B_1\rvert$ in R_1 and R_2.
4	$R_3 := C(R_3)$ $- C_8(R_2) \wedge C(R_1)$	$C_8(R_2)$ is the most significant bit of R_2. This is ANDed with $C(R_1)$, and the result is added to R_3. Sum is left in R_3, which must be a master-slave register.
5	$R_1 := 2R_1$ $R_2 := 2^{-1}R_2$	Shift R_1 to left ($\times 2$). Shift R_2 to right ($\div 2$).
6	As (4)	$C_8(R_2)$ is now second most significant bit of $\lvert X_1 - B_1\rvert$.
7	As (5)	As (5).
8 to 18	—	Operations as in (4) and (5) are applied alternately.
—	—	After step (18) R_3 contains $F - \lvert X_1 - B_1\rvert^2$.
19	$R_1 := \lvert X_2 - B_2\rvert$ $R_2 := \lvert X_2 - B_2\rvert$	As (3), except that X_2 and B_2 are sampled
20 to 34	—	As (4) to (17).
—	—	After step (33), R_3 contains $F - \lvert X_1 - B_1\rvert^2 - \lvert X_2 - B_2\rvert^2$.
End after $16q + 2$	—	Unless the processor has been stopped beforehand, the register R_3 contains $F - \sum_{i=1}^{q} \lvert X_i - B_i\rvert^2$

The X_i and B_i are integers in the range

$$0 \leq X_i, \qquad B_i \leq +127$$

F is an integer in the range

$$0 \leq F \leq 2^{21} - 1$$

The number dimensions (q) must not exceed 32. The decision M is calculated as follows:

$$M = \begin{cases} -1; \text{ if the output goes negative at any time, except on steps 3, 19,} \\ \quad 35, \ldots, (15(q-1)+3) \\ +1; \text{ if the output never goes negative before the end of step} \\ \quad (16q+2) \text{ (steps 3, 19, \ldots are again ignored).} \end{cases}$$

it would be possible to time-share between classification and the following tasks, although it must be remembered that a microcomputer is slow and has a small store.

(a) data logging,
(b) experimental control,
(c) data normalisation,
(d) statistical analysis.

Table 6.2 Summary of a digital circuit to implement a spherical classifier. After P. G. Ellis.

Integrated Circuits	Texas series 74 TTL logic
Number of dimensions	3
Clock frequency (max)	3 MHz
Time to calculate a decision	14 μs
Cost	£35*
Precision of X_i	8 bits

* The cost quoted here does not include the registers for F, $B_1 - B_3$, $X_1 - X_3$.

Table 6.3 Summary of a fully parallel digital circuit to implement a hyperspherical subclassifier. After M. V. R. Allin.

Integrated circuits	Texas series 74 TTL logic
Number of dimensions	16
Propagation delay	4 μs
Cost per dimension	£288*
Precision of X_i	6 bits

* The cost quoted here does not include the registers for F, $B_1 - B_{16}$, $X_1 - X_{16}$.

The microcomputer is also cheap enough for a large system to be built, which incorporates several of them.

The development of microcomputer and associated hardware (floppy discs, cassette tape recorders) is an advance. It is, however, too soon to realise the full impact they will make on pattern recognition.

6.8 IMPLEMENTATION OF THE LEARNING RULES FOR A CC

The growing and pruning rules may both be partitioned into quite distinct procedures:

(a) locate adjustment rules,
(b) weight adjustment rules,
(c) subclassifier selection (during learning),
(d) scoring,
(e) deletion/addition of subclassifiers.

Each of these may be implemented quite easily in electronic hardware ((a–c) are considered in [1]). However, the overall complexity of the growing and pruning rules is so great that there seems to be little purpose in trying to implement them completely in special-purpose electronics. We suggest, therefore, that a general-purpose digital computer be used, for at least part of the task. A hybrid system is shown in

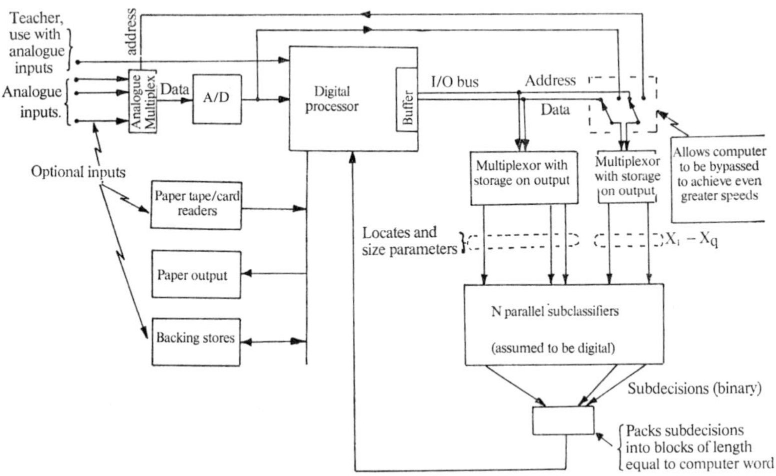

Figure 6.19 Hybrid system for pattern classification and learning. This system has inputs from paper tape, punched cards, magnetic tape/disc or analogue electrical channels. It has the full flexibility of the digital machine plus the great speed of the special-purpose hardware.

Figure 6.19 [3]. This offers a very flexible pattern classification facility, which can accept inputs from any of the following media:

(a) electrical, analogue,
(b) electrical, digital,
(c) paper tape, punched cards,
(d) teletype keyboard,
(e) backing store.

The bank of subclassifiers, plus the associated interface units form a hardware subroutine, which is many times faster than the equivalent software. If this subroutine is fast enough, we may transfer data between the processor and the hardware using processor-controlled I/0 without delaying the processor in a waiting loop. In most computers, it would be possible to transfer the complete **X**-vector in $3q$ instructions. The subdecisions would then be available from the $(3q + 1)$st instruction. Even greater speeds can be achieved, if data flows from the input medium to the subclassifiers, without the intervention of the processor. A path of this kind is shown at the top of Figure 6.19.

The locates and weights are held in both the main store of the processor and the multiplexor/store. This is an important feature since we should otherwise need to transfer the stored parameters from the multiplexor/store to core, and back again, during each modification. There is yet another way in which we can increase the speed. This is

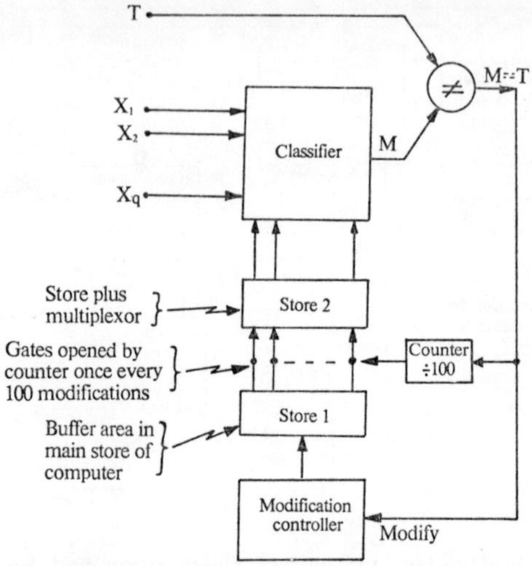

Figure 6.20 Illustrating the use of a buffer store to reduce the bandwidth of the I/0 bus, connecting the digital processor and the pattern classifier in the preceding figure. Store 1 represents the buffer area in the main store of the digital processor. Store 2 represents the hardware store which follows the multiplexor. After every 100 modifications store 1 is dumped into store 2.

simply accomplished by updating the contents of the multiplexor/store infrequently (Figure 6.20). We modify the "locates" and "weights" stored in core, and then after, say, 100 or 1000 modifications dump these new values into the multiplexor/store. The subdecisions being calculated are, of course, functions of the contents of the multiplexor/store. This modification to the learning rules has very little effect on their performance (Figure 6.21).

6.9 SUMMARY

In this chapter we have concentrated on practical details. We have described how the best results may be obtained, when using the CC training rules, described in the previous chapter. We have also pointed out certain pitfalls, and described how to avoid them. Appendix 1 summarises these ideas in a flow-chart.

The latter part of the chapter has concentrated on the implementation of a pattern classifier and learning techniques. We have given circuits for implementing a CC in analogue hardware, in a serial-parallel digital system, and in parallel combinational logic. The last of these, though the most expensive is potentially the fastest. Finally, we have described a flexible hybrid system, which can implement the growing and pruning rules.

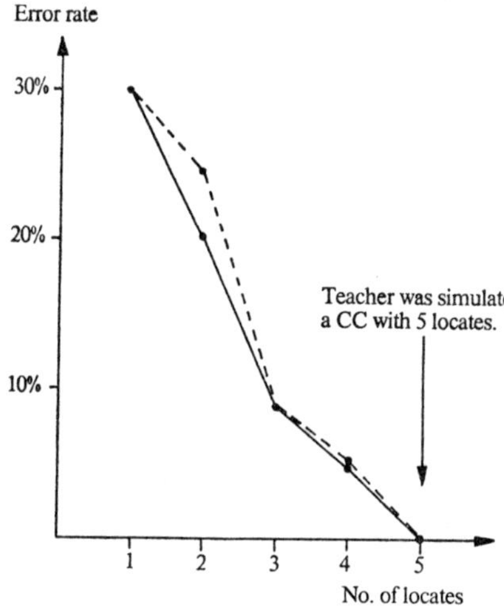

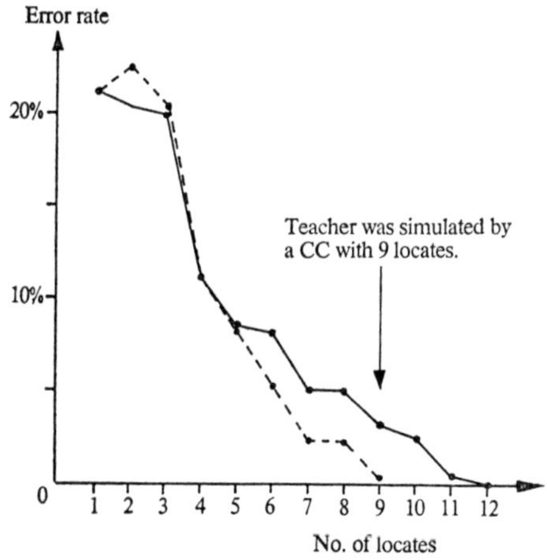

Figure 6.21 Comparison of growing rules. The solid curves refer to the normal growing rules, defined in Figure 5.31. The broken curves refer to the modified rules which account for the delay in the feedback path (see Figures 6.19 and 6.20). (A) Type A experiment. (B) Type A experiment, same data as Figure 5.37.

REFERENCES

1. B. G. Batchelor, Learning Machines for Pattern Recognition, *Ph.D. Thesis*, Southampton, 1969.
2. W. J. Conover, Practical Non-parametric Statistics (Wiley, New York, 1971).
3. B. G. Batchelor, Hybrid Pattern Classifier for Computer Input, in Computer Science and Technology, *I.E.E. Conf. Pub.*, No. 55, 141–150, 1969.

Chapter 7

Applications: Accomplishments and Proposals

7.1 INTRODUCTION

One of the factors which makes pattern classification so exciting, is the wide variety of applications that have been found. In this chapter, we shall describe some of these, and offer a few suggestions of our own.

The data analyst can only contribute to other subjects, if there is a complete rapport between himself and specialists in these subjects. This implies that both partners respect each other's judgement in their respective fields. Thus, a data analyst should avoid thinking that he can contribute directly to a field such as medicine. On the other hand, a doctor should realise that data analysis is a specialised topic, best understood by workers who have concentrated on it for many years. There are many difficulties to be overcome when analysing data, and even the most lucid book is a poor substitute for co-operation with an experienced analyst.

Our objectives in this chapter may be summarised as follows:

(a) To demonstrate the variety and range of applications, that have been suggested in the literature.
(b) To convince the reader that pattern classification is a useful addition to the repertoire of the data analyst.
(c) To suggest new applications, as far as these are apparent to the author.
(d) To illustrate some of the pitfalls which we have learned to avoid.
(e) To trigger the imagination of engineers, doctors, biologists and workers in other fields to find new ways of using pattern classification techniques.

The author has attempted to collect applications together, so that readers with one specialisation may find all that interests them in one or two sections. This has meant that proposals and well-tried applications are listed alongside each other. To avoid confusion all of the author's own suggestions are labelled *proposal.*

The author would like to explain his position. He has received no formal training outside that of an electrical engineer, and in particular has no authority to comment on medical matters. However, he has enjoyed the co-operation of a number of doctors in the work described in section 7.3, and the author is indebted to them for verifying the detail about these applications. Some of the other (non-medical) applications are also outside the author's range of experience. While he has made every effort to establish both the accuracy and the usefulness of his suggestions, the author realises that he may unwittingly offer solutions to trivial problems. The author has no desire to encroach on the activities of specialists in other fields, but feels he has a responsibility to describe as many possible uses for pattern classification as he can.

7.2 ENGINEERING

7.2.1 Optimal Control

(i) In some control systems a very complex controller is required. It may, in some situations, be cheaper to realise this, using a number of simple subsystems, which can be automatically selected as the system parameters change (Figure 7.1). Switching between control modes

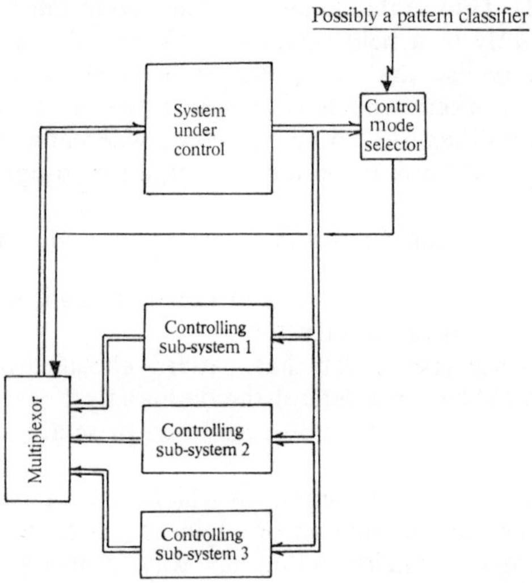

Figure 7.1 Construction of a complex control system by selecting the most appropriate sub-system. Selection could be accomplished by a classifier, whose inputs are derived from the system being controlled.

may be accomplished using a pattern classifier. Lucas [1] describes an example of this kind in his study of the control of a ship, during docking.

(ii) A classifier may be used in a bang-bang servo-system to control a two-state valve or switch.

Widrow and Smith [2] describe the use of a linear classifier to balance an inverted pendulum. While this application has little direct value to practical problems, it does demonstrate the basic principles in an effective way.

7.2.2 Modelling (Proposal)

In the preceding section on optimal control, we described how a complex control unit might be synthesised from a number of smaller subsystems. A pattern classifier could be used, in just the same way, to simulate a complex system using simple sub-models. The classifier would choose which of the simple models would be most appropriate for any given set of inputs (Figure 7.2).

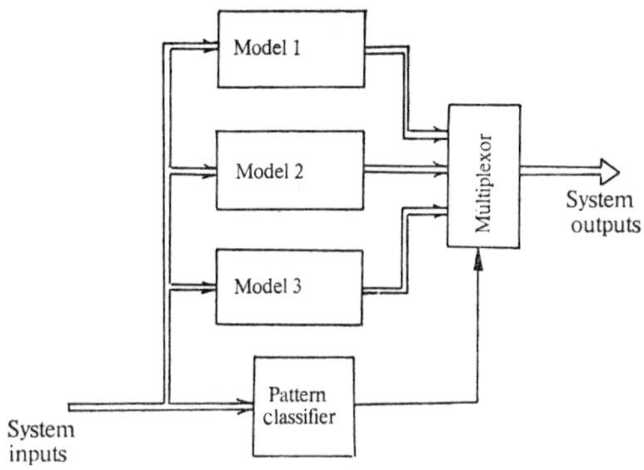

Figure 7.2 A complex simulation model may be produced by a series of simpler models, which are selected by a pattern classifier.

7.2.3 Hazard Warning (Proposal)

We might arrange for a pattern classifier to monitor a large plant and warn when it is approaching a dangerous state. The classifier could either initiate direct action, by some protective equipment, or merely notify human inspectors. The inputs to the classifier would depend upon the type of plant being protected.

There is a problem inherent in this application: how can we obtain measurements on typically dangerous conditions? Clearly, we cannot allow a real plant to approach a dangerous state. It may be possible to obtain a training set by using a simulation model, although it is doubtful whether this would be worthwhile, in a problem which is so well understood as to allow a complete simulation.

We shall discuss a similar problem (detecting pollutants in water) in Chapter 8, where we point out that this type of application requires a completely novel approach.

7.2.4 Quality Control (Proposal)

Three applications are suggested in this area.

(i) *Food production*

The senses of smell and taste, are perhaps the most difficult to define objectively. Of course the taste of, say, coffee may be attributed to the unique mixture of chemicals that this contains. If we could measure the concentrations of these substances, we could presumably define the limits of "coffee flavour". However these measurements would be difficult, since the chemicals contained in coffee are complex. We could measure other quantities, such as those listed below, rather more easily:

concentrations of certain ions,
pH,
optical absorption, at various frequencies,
conductivity,
osmotic pressure, compared to certain standard solutions
specific gravity,
viscosity,
optical activity.

It might be possible to automate the measurement of these items to obtain the inputs to a classifier.

Suppose we now prepare a large number of beverages (which are all reputed to be coffee) and ask a human subject to label them as "acceptable coffee flavour" or "not acceptable coffee flavour". Assuming that our suggested measurements are relevant,* we now have a set of data for training a classifier to distinguish between acceptable and non-acceptable flavours. The use of a classifier to monitor flavour has two advantages apparent:

(a) It would provide an automatic method for monitoring the flavour of food during processing.
(b) It would provide an absolute standard for "assaying" new products.

The second of these is a much more difficult problem than (a), since many of the measurements we have suggested are only loosely connected to the chemical concentrations we should really like to measure.

(ii) *Colour recognition*

A similar application may be foreseen in the dye, paint and textile industries, where a pattern classifier could be used to control the colours of dyes. The descriptors would presumably represent spectral components of light reflected from the surface being inspected.

* Without the advice of a specialist in food technology we cannot be sure that this set of descriptors is adequate.

(iii) *Chemical analysis*

Pattern recognition may be useful in other aspects of chemical engineering to provide automated analyses in conjunction with gas-spectrometer, or chromatographic equipment. Would it be possible to specify smells, as we have attempted to define the sense of taste?

7.2.5 Fault Diagnosis

To illustrate the diagnosis of faults in electrical equipment, we shall consider an analogue summer, with two inputs. A simple test circuit is shown in Figure 7.3A. Two measurements are derived (X_1, X_2), which

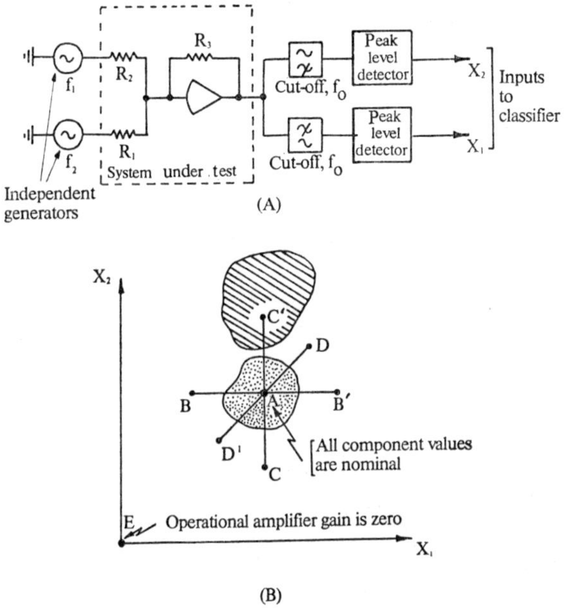

Figure 7.3 A simple electrical test system incorporating a classifier. (A) Test circuit. (B) Measurement space. Stippled area represents an acceptable circuit. The cross-hatched region corresponds to an unduly low value for R_2.

we may use to draw a descriptor space (Figure 7.3B). Variations of the components will cause the point (X_1, X_2) to move. For example, changes in R_1 will cause the point (X_1, X_2) to move along the line BAB′. Hence, to recognise a summer that is within specification, we need only check that (X_1, X_2) lies within the stippled region. Points within the cross-hatched region are typical of those in which R_2 is low.

A more complex system might require more test signals, (not necessarily sinusoidal, as in Figure 7.3). In Figure 7.4 we show a possible method of testing the frequency response of a filter or amplifier.

Levadi [3] discusses fault diagnosis in electrical equipment at greater

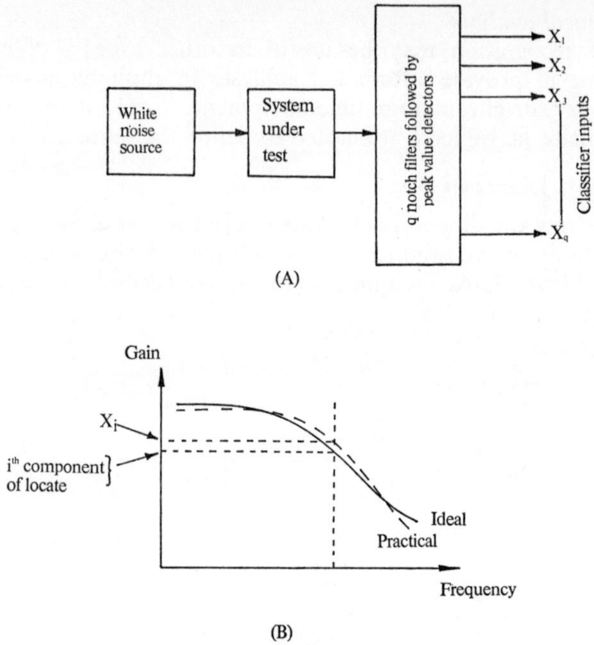

Figure 7.4 An electrical test system in which the descriptors are spectral ordinates. (A) Test circuit. (B) Relationship of the descriptors and locates to the spectrum.

length. Cox and Renken [4] consider the use of pattern recognition methods for fault-finding in other fields of engineering.

7.2.6 Prediction (Proposal)

A pattern classifier is not particularly well suited for prediction, because its output is digital. Many attempts have been made to use a classifier as a predictor, but they either result in a poor resolution for the predicted variable, or an excessively large number of classes. For example, an attempt was made to predict the national demand for electrical power [5]. This problem required that the power demand be predicted with an error of less than 5% of the maximum possible load. This resulted in a 30 category classifier! Other prediction techniques produced better results than the classifier.

An alternative arrangement might be more successful, although the method is so far untried. This technique predicts a range (possibly not continuous) of values of a variable (Figure 7.5). The classifier has inputs equal to:

$(q-1)$ independent variables $(X_1, \ldots, X_{q-1})$ representing the inputs to the predictor.

X_q which is equal to S, the predicted variable.

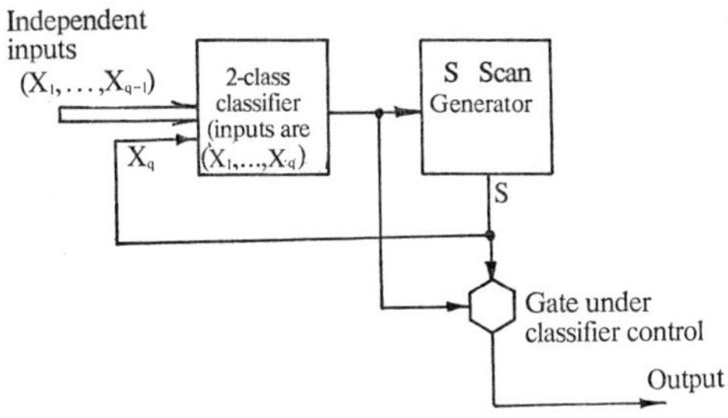

Figure 7.5 A predictor which uses a pattern classifier to detect acceptable values. This system produces a range of acceptable values rather than a single numeric quantity. This range may not be continuous.

The classifier output indicates whether S is an acceptable prediction, given the inputs $(X_1, \ldots, X_{q-1})$. The device labelled "S-scan generator" varies S, possibly in response to the classifier output. If the classifier indicates that S is an acceptable prediction, the gate is opened, and S appears at the output. The only advantage offered by this system, appears to be its ability to present a range of values at the output, rather than a single numeric quantity.

7.3 MEDICINE

Many authors have realised the potential value of applying pattern classification methods to medicine, and a multitude of examples have been described in the literature [6–15]. Diagnosis seems to be the most popular application, although it is possible to find other rôles for the classifier, including:

(i) prognosis,
(ii) choosing therapy,
(iii) allocating resources to patients,
(iv) on-line administration of drugs,
(v) patient monitoring,
(vi) epidemiology.

Provided that a suitable analysis unit exists,* a pattern classifier may be applied to virtually any problem in decision-making. Hence we can expect to find applications in such diverse fields as EEG analysis, X-ray interpretation, psychiatry, chemical pathology. The reader will notice that these four examples all demand different types of input to the analyser:

(1) electrical (EEG, ECG, EMG),
(2) visual (histology, blood cell counting),
(3) paper (psychiatry, epidemiology),
(4) chemical (pathology).

* This is always a major assumption, but is one that the author is forced to make, since he cannot discuss medical points adequately.

In some applications, we may find that we need to accept data from more than one of these media. It is especially important to notice the third of these (paper). It is often much easier to design a proforma, or questionnaire, than to make judgements from it when it is complete [6]. We can therefore expect to see a number of applications in which a doctor completes a proforma, which he has designed, and then hands it to an analyst for interpretation.

In medicine it is particularly important to establish a protocol. The doctor will use his skills in two ways:

(a) to design the proforma or select a set of descriptors,
(b) to provide a training set for the classifier.

The classifier can offer several valuable features, including:

(a) the ability to combine the opinions of several doctors (the classifier is trained using a consensus),
(b) consistency, without fatigue,
(c) greater speed than a human being,
(d) reduced cost/decision, compared to a doctor.

The disadvantages of mechanical decision-making are obvious to both doctors and patients. It is important that we do not forget:

(a) the distrust of the patient for mechanical devices,
(b) the greater isolation of the doctor from the decision,
(c) the inflexibility of the classifier, compared to the doctor.

Table 7.1 List of items supplied to diagnose cancer

Item	
Age	
Height	
Weight	
Daily urinary output	
No. of pregnancies	{Live births {Still births
Hormone metabolites (concentrations in the blood)	{11-Deoxy-17 oxosteroids {17-Oxogenic steroids
Patient's estimate of total time she has spent breast feeding	

Each item was measured by a single number, which was rescaled to lie within the same range.

7.3.1 Diagnosis of Breast Cancer

Data on 200 women was available* (Tables 7.1 and 7.2 summarise its format). We analysed this data, using the technique described in section

Table 7.2 Showing numbers of women in the sample studied for diagnosis of breast cancer

Category		Number	Method of diagnosis
Collectively labelled "cancer" (130 women)	Stage I	35	Histological examination (Stages I–II)
	Stage II	51	
	Stage III	29	Clinical examination (Stages III–IV)
	Stage IV	15	
No cancer		70	Women who were not ill, all ages, but not children

5.4.1 and obtained the results summarised in Table 7.3. We also applied a CC to this problem, and obtained the results shown in Table 7.4.

We deduced that there is considerable overlap between the classes,

Table 7.3 Results obtained using the method of section 5.4.1 to analyse the data on breast cancer

	Predicted	
	No cancer	Cancer
No cancer (70 women)	32	38
Cancer (130 women)	44	86

but that despite this, we had probably obtained nearly optimal results. We must emphasise however, that before we can place great weight on this conclusion, we should test the CC using some new data.

Table 7.4 Results obtained with a CC applied to the data on breast cancer

	Predicted	
	No cancer	Cancer
No cancer (70 women)	49	21
Cancer (130 women)	26	104

Three subclassifiers were used. (The CC was trained, and these results obtained using all of the patients. It would have been better to divide the data into two groups of approximately equal size, train the CC on one group and find the error rates using the other.)

* Supplied by Dr. H. Miller.

7.3.2 Choosing Therapy for Peptic Ulcers

Of those people who suffer from peptic ulcers, about 5% will not respond satisfactorily to the operation called vagotomy. In these cases, the patients will usually be subjected to a second operation. Clearly it would save valuable resources, as well as the patient's trauma, if the first operation could be avoided in those patients who would not respond satisfactorily. The success of vagotomy may be judged by considering the level of acid secretion in the stomach. Success is usually assured, if the level of the stomach acid is reduced by 20% or more.

We were given data on 134 patients who had peptic ulcers.* Of these, 8 had acid level reductions below 20%. We were required to predict which patients would not respond to vagotomy, given 37 personal and biochemical descriptors.

It was possible to separate the two groups perfectly, although the sample was so small that we cannot be sure that this result is reliable. To achieve greater reliability, we require much more data, especially from the rarer class. This example illustrates a frequently encountered difficulty in automatic medical diagnosis, it is often very difficult to collect sufficient data.

7.3.3 Prognosis following cardiac arrest†

During cardio-respiratory arrest, the brain is deprived of oxygen. This sometimes results in the patient suffering irrecoverable brain damage which may lead to death, while in other cases he may fully recover. The EEG has been used to detect anoxic brain damage, and hence predict the outcome (necessary for proper case management). The conventional approach, based on intuitive design of prognostic rules, has only been partially successful. These rules cannot be applied in about 13% of the cases (Prior, p. 231). However, in those situations where the prognostic rules may be safely applied, there is a success rate of 80–85% (Prior, p. 248).

The EEG recording was manually coded using a proforma (Figure 7.6), to yield a 13-dimensional pattern vector. Two complementary CCs (section 6.2.7) were designed using a training set for which the outcome was reliably established as full recovery or death due to anoxic brain damage. Since then these classifiers have been applied to other clinical groups. (Figure 7.7) Dr. Prior (p. 231) states that when the classifiers are combined, "the number of records leading to erroneous conclusions are very few (of the order of 2–3%), and most derive from patients with barbiturate coma or mild anoxic damage which largely spared the

* Supplied by Dr. I. Glen, formerly with M.R.C., Graylingwell Hospital, Chichester.

† Rewritten just before publication, in the light of a book by Dr. P. F. Prior ("The EEG in acute cerebral anoxia", Excerpta Medica, 1973). I am grateful to both Dr. Prior and Mr. D. E. Maynard for their help in the preparation of this section and Figure 7.7.

FREQUENCY COMPOSITION

Alpha activity:	absent 1	left	1 2 3 4 5 6
	scattered waves only. 2		
	poorly sustained 3	right	1 2 3 4 5 6
	moderate amount 4		
	prominent. 5		
	large amount 6		
Fast activity:	absent 1	left	1 2 3 4
	scanty 2		
	moderate (normal) 3	right	1 2 3 4
	profuse (excessive) 4		
Intermediate slow activity:	absent 1	left	1 2 3 4 5 6
	scattered waves 2		
	moderate (normal) 3	right	1 2 3 4 5 6
	prominent (normal) 4		
	moderate excess 5		
	gross excess .6		

Figure 7.6 Proforma for assessment of EEGs following cardio-respiratory resuscitation. Only the first 6 items are shown. The rating scales are highly subjective and the annotation ("absent, scanty, moderate, etc") acts only as an aide memoire. (Reprinted by permission of the Electro-Physiological Technologists' Association.)

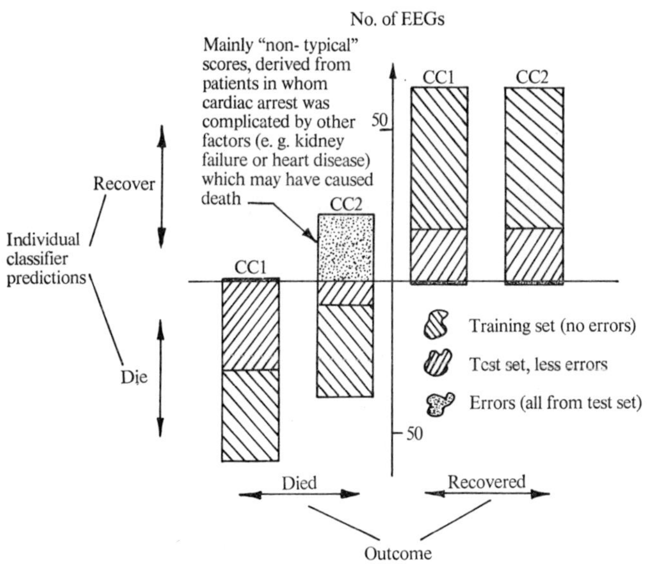

Figure 7.7 Results obtained after training two CCs (CC1 and CC2) to predict the outcome of cardio-respiratory arrest. CC1 was trained to enclose the class 'recover' and CC2 the class 'died'. Both classifiers were trained on a carefully selected set of patients in whom the cardiac arrest was uncomplicated. Predictions on the test set were based on the difference between individual scores (see section 8.5) obtained from CC1 and CC2.

cerebral cortex". Mr. D. E. Maynard has built a hardware device‡ to perform this prognosis. This is in regular use at the London Hospital.

7.3.4 Medical Screening (Proposal)

This is perhaps the area most likely to benefit from the use of pattern recognition techniques. The essential features of screening are that it must be cheap, and that it should "capture" as many ill people as possible, while keeping the number of "escapes" to a minimum. Screening may take place in several distinct stages. At each stage the decision is either:

the person is well and can be dismissed from further consideration

or

the person is possibly unwell and should be considered further.

The ideal multi-stage screening procedure uses a hierarchical decision structure (Figure 7.8). To minimise the overall cost of the screening

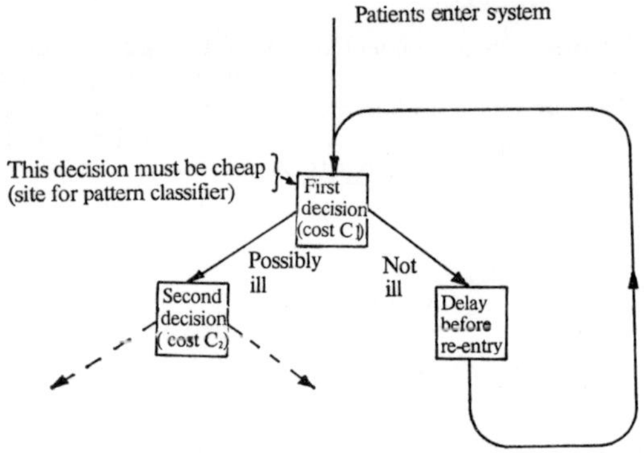

Figure 7.8 It may be quite acceptable to use a cheap low-accuracy, decision-maker to eliminate "safe" patients from further consideration. This would require that the decisions of the first stage be biased to produce fewer "escapes", at the expense of introducing more "false alarms". For the system to be effective, C_1 must be much smaller than C_2P_F, where P_F is the probability of the first stage producing a false alarm.

programme, the first decision-making device should be the cheapest (C_1 less than C_2 in Figure 7.8), and should be biased to give the decision "possible" in doubtful cases. This will reduce the number of "escapes" at the expense of increasing the "false alarm" rate.

There are enormous economic benefits to be obtained, if we can use a classifier to provide the first stage decision in a screening organisation. To do this, we must find some way of biasing the decisions, so that the decision "possible" is given in any doubtful cases. It is possible to do this only very crudely at the moment, for example, by expanding or contracting the circles in a CC (Figure 7.9). This problem is discussed again in Chapter 8, and Appendix 4.

‡ D. E. Maynard and P. F. Prior, Development of a prognostic Pattern Recognition Instrument as an aid to visual assessment of the EEG, Electro-enceph, clin. Neurophysiol., **34,** 108, 1973.

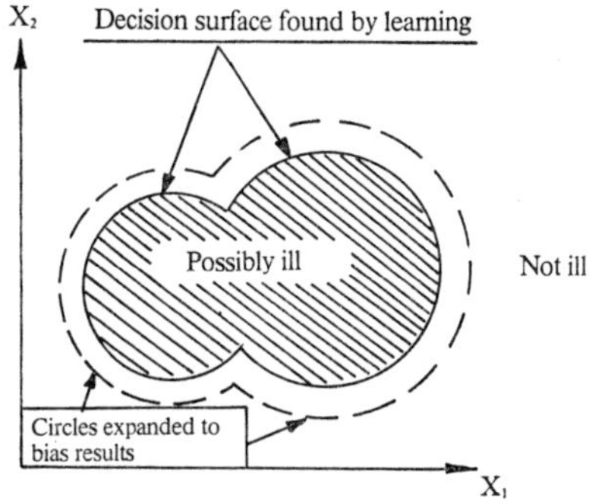

Figure 7.9 A crude method of biasing a compound classifier to increase its safety.

7.3.5 Epidemiology (Proposal)

The author* is currently studying the problem of diagnosing migraine using data, collected by questionnaire [17]. The importance of this project is that if successful, it will provide a method for analysing new questionnaires without further reference to a clinician. This would greatly reduce the cost of further surveys to study the incidence of migraine. It is too soon to be able to report any results.

7.4 VISUAL PATTERN RECOGNITION

Visual pattern recognition [18] is concerned with a wide variety of problems. Just a few of these are listed below.

7.4.1 Fingerprints

The problem of recognising fingerprints is quite different from almost all other applications, for the simple reason that a criminal will not make any special effort to leave good impressions. On the contrary, we can be sure that a diligent criminal will take every precaution to make our job more difficult!

Many of the proposed techniques for recognising fingerprints make use of processing techniques which gradually "clean up" a computer image of the print, and reduce it to a stylised form [19]. This forms the basis for the recognition process, which more closely resembles a syntactic analysis than a pattern classifier. Other techniques have been suggested, which measure the curvature of the ridges. This is rather more interesting to us, since the print can then be represented by numeric descriptors. However, a print left at the scene of a crime is of variable size, depending upon such factors as the position of the finger, type of surface and finger pressure. This means that the number of ridges in the scene-of-crime print varies, and that there is a variable number of descriptors, (see section 8.7). The difficulty of encoding

* In collaboration with Dr. W. E. Waters.

the print and the non-co-operation of the criminal, together make this a very unlikely area for the successful application of pattern classification.

7.4.2 Aerial Reconnaissance

Military and meteorological reconnaissance are usually concerned with the detection of complex images containing many "objects" of interest. The recognition of these objects might be possible using pattern classifiers, but the analysis of the whole scene would require other techniques, for example syntactic analysis. Thus, we might be able to detect individual lorries or buildings using classifiers, but would require other techniques to count them and decide on their significance.

Another problem that can be tackled using aerial photography is the estimation of crop distributions. From an aircraft, a field of wheat has a colour which enables us to distinguish it from, say, root vegetables. To estimate the area under cereals we should require the following descriptors, in addition to the colour measurements:

(a) time of year,
(b) light conditions near the ground (measurable by using the colour of roads or by direct measurement from low-flying aircraft).

We should also need to know what varieties of crops are likely to be encountered in the area. With these descriptors we should have the necessary information to perform a spot identification of crop types. The United States Government have been using aerial reconnaissance techniques for crop identification, for several years.

7.4.3 Medical Material

The screening of women for cervical cancer provides a good example of this type of application [20]. If this service were to be provided to all women in Britain over the age of 25, once every 5 years, about 3½ million tests would be required each year. To minimise the cost and increase the availability of the service, attempts have been made to automate the visual examination of smear slides. Similar objectives have motivated the efforts to automate the screening for chromosome abnormality [21]. X-ray interpretation also provides a field where automation might bring significantly reduced costs (22).

7.4.4 Optical Character Recognition

This is the longest established branch of our subject, and has to a great extent moulded the research. We have already seen how type-fonts vary (Figure 2.11). Even within a single font, there is a vast range of print quality, due to paper, typewriter-ribbon and type-face imperfections (Figure 7.10). Despite these variations it is possible to read single-font numerics to high accuracy. If the number of symbols is increased to include the alphabetic characters, the accuracy will be

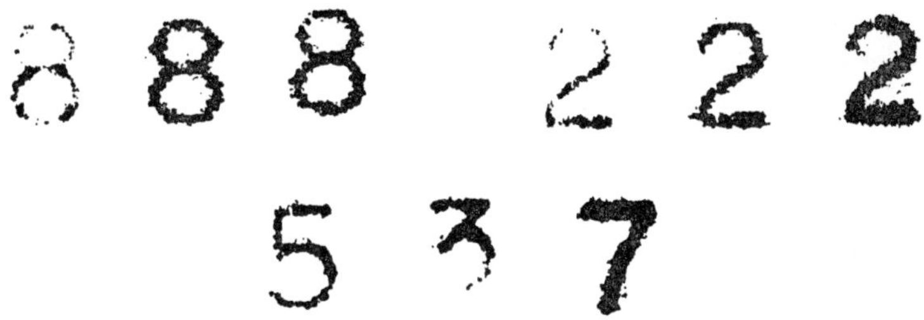

Figure 7.10 Photographs of some of the numerals used in the study of optical character recognition. Magnification ×5.
Supplied by Dr. J. R. Parks (National Physical Laboratory, Teddington).

reduced (Table 7.5). Increasing the number of fonts will generally have the same effect.

Carefully drawn numerals can also be recognised with a reasonable accuracy [2] (Table 7.6). An alternative approach to recognising hand-

Table 7.5 Summary of the performance of a fixed font optical character recognition machine. (After Funakubo, [23])

No. of fonts	1	IBM Advocate
No. of classes	16	0–9,C,S,T,X,Z,/
No. of samples	150	—
Average recognition rate	97.7%	—
Worst case recognition rate	90%	Number 8 often confused with 0 or 9.

writing is offered by sensing marks on a specially prepared form (Figure 7.11). Although the recognition equipment is cheaper, than for either typed or hand-written characters, the format is restricted by the printed form. It must also be remembered that the information density (bits/page) is lower for mark-sensing.

Table 7.6 Summary of the performance of a machine for recognising handwritten numerals. A piecewise linear classifier was used. (After Saraga, *et al.* [24].)

Data	No. of characters	Recognition rate
Training set (digits 0–9 drawn by 8 people)	400	100%
Test set I (drawn by same people)	400	96%
Test set II (drawn by 3 other people)	150	84.7%

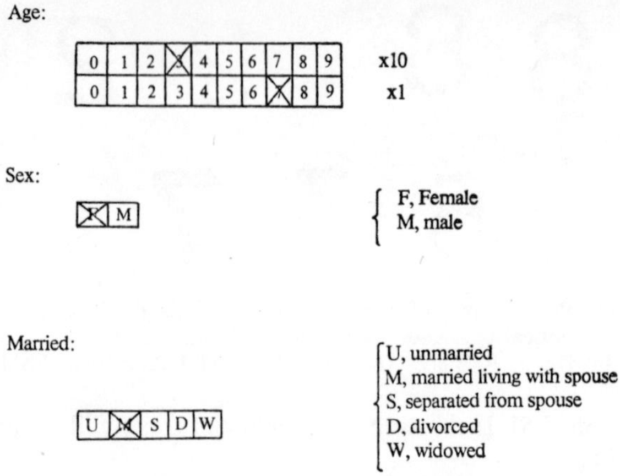

Figure 7.11 Part of a form used in mark-sensing, showing the coding for a married woman, aged 37. Notice that we apparently *delete* the answer we wish to give.

It seems appropriate at this point to describe our own results on an optical character recognition problem. The data* is summarised in Table 7.7. The print quality may be judged from Figure 7.10. Table 7.8 summarises our results. Here, we have used brackets to indicate the class label. For example, (8) is a type-written number "8". (0, 1, 2, 3, 4) is a class containing the numerals "0–4". We use 3 × (4) to denote 3 examples of the typed number (4). Rows 1–10 of this table show that we could construct a bank of 10 parallel CCs, to recognise the numerals.

Table 7.7 Summary of the data for optical character recognition

No. of samples/class	96
No. of classes (digits 0–9)	10
No. of descriptors	20
Resolution of descriptors	64 levels
Font	ICT
Height of type	0.1″
Printer	ICT 609 line printer
Paper reflectance	80%
Ribbon type	silk
Ribbon quality	24 samples, new
	24 samples, medium age
	24 samples, old ribbon
	24 samples, stationary

For further details see [25, 26].

* Supplied by Dr. J. R. Parks (National Physical Laboratory Teddington).

Table 7.8 Results obtained using a CC to recognise typed numerals

Set S, indicating numeral(s) to be enclosed inside the hyperspheres	Errors: Confusion of members of $\bar{S}$ for S	Errors: Confusion of members of S for $\bar{S}$	Comments
(1)	—	—	5 hypersphere subclassifiers were used, although only one was effective (other 4 were degenerate). One of the initial locates was a crude estimate of the mean vector for S. Other 4 were randomly scattered in the space.
(2)	—	5 × (2)	
(3)	—	3 × (3)	
(4)	2 × (8)	3 × (4)	
(5)	—	5 × (5)	
(6)	1 × (2)	3 × (6)	
(7)	—	—	
(8)	1 × (5)	5 × (8)	
(9)	—	5 × (9)	
(0)	—	—	
(1, 3, 5, 7, 9)	4 × (8)	2 × (3), 6 × (5), 6 × (9)	5 hyperspheres, all effective. Initial locates found by estimating mean vectors for those numerals in S.
(0, 1, 2, 3, 4)	2 × (5), 4 × (8)	6 × (2), 2 × (3), 4 × (4)	
(0, 1, 2, 4, 7)	7 × (2), 5 × (4)	2 × (8)	Initial locates found by maximindist (see Appendix 2)

The data consisted of 96 samples of each digit. The classifier was trained and tested on the same data. (Based on data supplied by Dr. J. R. Parks.)

This would have an error rate of about 0.4% and a reject* rate ("don't know" category) of 3%.

In retrospect the author can criticise this work in a number of ways:

(a) He did not allow enough time for learning.
(b) k was rather large (0.05).
(c) There was not enough data.
(d) There was no estimate of the Bayes' error probability.
(e) There was no independent check on the error rate using a test set.
(f) We did not use a good method for selecting the initial locates.

Rows 11–13 are interesting, because we were able to separate multi-cluster classes quite adequately. These results were our earliest evidence that the CC learning rules can adequately cope with multi-cluster classes in a high-dimensional space (20-space).

7.4.5 Other Applications

The following list, although only a small sample of the proposed applications, should convince the reader of the broad range of activities that are being studied under the title of visual pattern recognition.

* A reject occurs when none of the 10 CCs produce an output of +1.

(a) graphology [27],
(b) taxonomy [28],
(c) vehicle/aircraft recognition,
(d) bubble/spark chamber photographs [29],
(e) geological samples [30, 31],
(f) histology,
(g) automatic visual inspection, for quality control.

The author can foresee applications in other areas:

(h) bank note/coin recognition for bank automation,
(i) harvesting; recognising ripe fruit for example,
(j) archeology; matching new finds to existing knowledge,
(k) ecology, automatic recognition of plant/insect types for finding, distributions.

7.5 TIME-SERIES ANALYSIS

As in the previous section we can list a large number of applications under this title:

(a) speaker recognition, useful as a prelude to speech recognition [33],
(b) speech recognition (section 7.5.1),
(c) bird-song recognition (section 7.5.2),
(d) ECG, EMG and EEG (section 7.5.3),
(e) radar and sonar target-recognition,
(f) vehicle recognition (section 7.5.4),
(g) intruder detection using Dippler-radar.

7.5.1 Speech Recognition

Speech recognition is widely acknowledged to be one of the most difficult problems in our subject [34–36]. This opinion has been developed over a number of years as continuing research effort has resulted in little improvement in the capabilities of the machines. A typical research programme has sought to construct, or simulate, a machine with a repertoire of 12–20 words, and which can cope with many different speakers. It is usual to find that the machine performs well with those speakers who provided data for the training set, but very badly with others. Another frequently occurring phenomenon is the unconscience learning by the speakers to produce the desired machine output.

Despite this unhappy situation there are several ways in which the usefulness of these machines could be extended:

(a) Using a *speaker recognition* device, the optimum *speech* recognition strategy could be selected, before we attempt to perform any recognition of words (see section 7.2.1). Such an approach would probably demand that a standard text be recited, when a new speaker uses the machine.*

(b) A simpler approach could be developed if we asked the new speaker to set a switch, to indicate his identity to the machine, before he speaks. This would change the parameters of the classifier. Of course, the classifier would have to be trained for each speaker.

(c) We could deliberately train each new user to obtain a good performance. This might be tolerated by people who are disabled, and are strongly motivated to develop an effective speech communication path to a computer [37].

(a) and (b) would require that each new customer provides data for optimising the machine to suit his accent. This should not be impossible, but it would clearly be rather expensive to collect and analyse sufficiently large amounts of data.

7.5.2 Bird-Song Recognition

The recognition of bird-songs is important for two reasons:

(a) to identify birds near an airport, so that an appropriate recording can be played to frighten them, and clear the air space,

(b) for ecological surveys.

One of the author's colleagues† is currently studying this problem using spectral analysis. It is too early to report any results.

7.5.3 Electroencephalography

We have already described one application of pattern classification techniques to the EEG (section 7.3.3), where we made use of a proforma completed by a human observer. In this section, we describe two further projects which have used fully automatic techniques for describing the EEG. In both cases the preliminary analysis was performed by a computer program simulating the EEG measurement techniques suggested by Hjorth [38].

7.5.3.1. Recognition of Alert and Drowsy States

During an EEG recording session, the subject, an epileptic boy, was allowed to become drowsy. After about 40 minutes he was suddenly roused by a man entering the recording room. We were able to distinguish perfectly between the drowsy and alert states (before and after entry), using the EEG.

Of course, we considered only one patient in this study, and it would be dangerous to assume that we could extend the technique to other subjects. Other authors have successfully used pattern recognition techniques for distinguishing between various levels of sleep [39].

* This is the basis of a machine produced by International Computers Limited.

† Dr. D. W. Thomas is studying this problem in collaboration with Dr. P. E. Howse, Department of Biology, University of Southampton.

7.5.3.2 *Recognition of Paroxysmal Activity in the EEG*

Data on 6 patients, with temporal-lobe epilepsy, was collected* and coded as follows:

(a) A single-channel EEG was recorded on a digital tape-recorder.
(b) The 3 Hjorth parameters were calculated for each epoch in the record (about 240 epochs/patient, epoch length 1.28 sec).
(c) 7 paper charts were drawn, each containing the unprocessed EEGs, for all 6 patients (in sequence). In addition to the EEG trace, there was a time-marker indicating the limits of each epoch.
(d) 7 EEG observers, with experience ranging from 1–20 years, then coded each epoch, stating whether or not they thought it contained any paroxysmal activity.†

It is interesting to compare the ability of human observers (see Table 7.9). Observers 6 and 7 are the least experienced, while observers 1 and 3 are the most experienced. Observer 3 is far more likely to indicate paroxysmal activity, than are any of his colleagues. This is the most senior observer, so his decision will be taken as the standard.

Table 7.9 Comparing the decisions of 7 EEG observers of varying experience.

		Observer giving "no paroxysm" decision						
		1	2	3	4	5	6	7
Observer giving "paroxysm" decision	1	—	10.9	6.9	13.1	14.8	24.3	23.1
	2	6.4	—	4.2	9.2	10.9	19.9	18.9
	3	20.7	23.5	—	26.2	28.0	38.8	37.7
	4	4.7	5.2	2.9	—	8.7	15.5	15.5
	5	3.3	3.8	1.5	5.6	—	12.9	12.3
	6	0.6	0.7	0.3	0.3	0.8	—	2.2
	7	0.6	0.8	0.3	1.3	1.3	3.4	—

The entry in column *i*, row *j* indicates the percentage of instances when observer i thought there was no paroxysmal activity, while observer j thought there was. Observers 1 and 3 are the most experienced. (Data supplied by Dr. C. D. Binnie.)

In Table 7.10 we compare the human observers on the one hand, and the fully automated recognition equipment on the other. The machine (the Hjorth analysis plus the classifier C_1) compares favourably with the human observers on the *paroxysm* class. However, on the *no-paroxysm* class the error rate of C_1 is larger than the corresponding figure for a human observer.

* Dr. C. D. Binnie, Department of Clinical Neurophysiology, St. Bartholomew's Hospital, London.

† Paroxysmal activity may be regarded as any EEG activity that is not typical of the normal subject. Notice that paroxysmal activity may last for much less than the epoch length. This makes the recognition of paroxysmal activity more difficult, than if it lasted for the whole epoch. We return to this point in Chapter 8.

Table 7.10 Comparing the ability of human observers and pattern classifiers to recognise paroxysmal activity in the EEG

	Observer						Classifier	
	1	2	4	5	6	7	C_1	C_2
Observer 3 indicates "paroxysm" (48.9%)	44.5	48.0	53.6	57.1	79.1	77.3	25.6	40.3
Observer 3 indicates "no paroxysm" (51.1%)	13.5	8.2	5.7	3.0	0.5	0.7	32.4	36.4
Average	28.6	27.6	29.2	29.5	38.7	37.8	29.1	39.5

Observer 3 was the teacher and his definition of paroxysmal activity is taken as the standard. Entries in the tables are percentage disagreements, compared to observer 3. Thus on 44.5% of those occasions when observer 3 indicated paroxysmal activity, observer 1 disagreed and indicated there was no paroxysmal activity. C_1 was a CC with 2 subclassifiers (paroxysm class inside spheres). C_2 had 3 subclassifiers (no paroxysm class inside spheres). (Based on data supplied by Dr. C. D. Binnie.)

At the time of writing, this work is incomplete. Our next task will be to bias the classifier to vary the error rates in the two classes (see Chapter 8 and Appendix 4).

7.5.4 Vehicle Recognition

The ability to recognise a vehicle, from the sounds emitted by its engine, would be useful in military surveillance and traffic census-taking. One of the most difficult problems is the determination of the optimum sampling period for the analysis equipment. If this is not carefully chosen to coincide with an integer multiple of the engine firing rate, the descriptor vector is very variable. An important step has been taken by D. W. Thomas who has developed techniques for calculating the firing rate [41]. Moments of the acoustic waveform have been proposed as descriptors (Figure 7.12).

7.6 CLASSIFICATION OF BIOLOGICAL DATA

We have already briefly discussed the use of pattern classification techniques in biology (see section 2.7.1, and in particular Figure 2.13). Further examples of descriptor spaces may be found in Sokal and Sneath's book on numerical taxonomy [42]. However, they regard a *group* of animals or plants as a unit* and perform their analysis on these units. Sokal and Sneath do not use individual organisms because

* A unit may be a race, species, genus or any other (useful) group. The unit is usually represented by its mean vector.

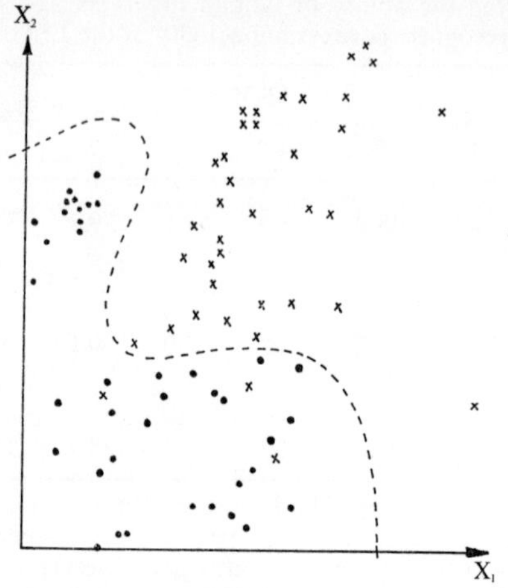

Figure 7.12 Showing the distribution of pattern points for the recognition of vehicles using acoustic signals. X_1 and X_2 are moments of the acoustic wave-form. Crosses correspond to vehicles with a Ford 4D diesel engine, and circles to vehicles with a Perkins 3 diesel engine. In both cases there are three different vehicle-body types, with the same engine. The vehicles were stationary, and the engines were ticking over while these measurements were taken. The broken line indicates a possible surface for classifying the wave-forms according to their origin. (Reproduced by permission of Dr. D. W. Thomas, SOURCE [40].)

they regard the analysis of large data sets as totally impractical. This is certainly not the case now; with the development of cheap, digital electronics it would be feasible to construct special-purpose machines to analyse much larger sets of data than they envisaged. The use of electronic classifiers would be of the greatest value in ecological surveys, where there is an enormous amount of data waiting to be processed. However, as so often happens, we must await the development of automatic pattern description techniques, before we can see the full benefits of our work.

7.7 LITERATURE STUDIES AND INFORMATION RETRIEVAL

Suppose that we have been given two samples of English prose by different and known authors. How can we decide who is more likely to have written a third document, whose authorship is in dispute? One possible solution would be to count the number of occurrences of certain words, or pairs of words, and use these numbers as descriptors. Following this approach, except that we used letter, not word frequencies, we were able to separate English, Latin, French and German prose* [26, 43]. We can see only two practical applications for language recognition:

(a) Automatic optimisation of coding for ciphers. A new cede would be selected depending on the source language.
(b) Optimisation of coding for telecommunications.

Pattern recognition techniques have also been considered for use in an information retrieval system. The efficiency of the system would depend upon the ability to describe documents in an effective way. Our techniques require numeric descriptors (or logical entities) and equate distance in the descriptor space, with dissimilarity. Needham [44] discusses a coding scheme, which makes use of this equality, and which uses numeric descriptors.

The author's suggestion for an information retrieval system, based on pattern classification techniques, is outlined in Figure 7.13. A useful

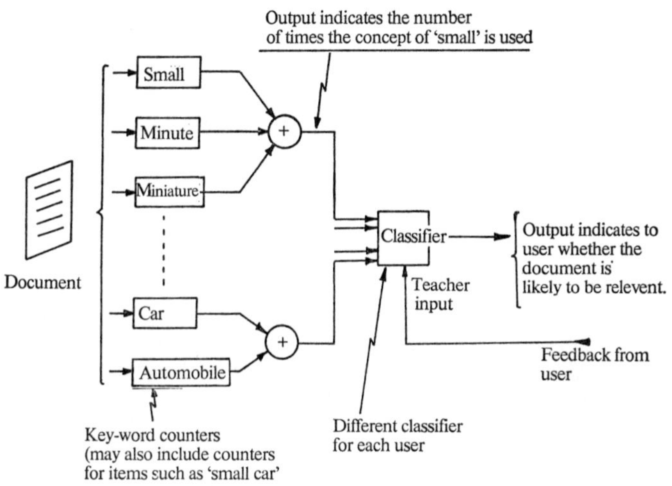

Figure 7.13 An information retrieval system based on a pattern classifier. All equipment is common, except that a new set of classifier parameters are used, for each new user. Notice the feedback path from the user to the classifier.

feature of this system is that the user could improve his service, by training the classifier via the feedback path.

* Natural languages have proved to be a very convenient source of data for studying pattern classification in high-dimensional spaces. We can adjust the overlap between classes, by varying the sample length. We can form classes such as (English *or* French) to generate multi-modal distributions. We can also vary the number of descriptors over a wide range (typically 2–26). The use of natural data also adds the conviction that there is no cheating.

REFERENCES

1. K. Lucas. "An approach to unsupervised learning in goal-seeking control systems", *Ph.D. Thesis* University of Southampton, 1974.
2. B. Widrow and F. W. Smith, Pattern Recognising Control Systems, in Computer and Information Sciences, 288–317, Spartan, Washington, 1964.
3. V. S. Levadi, Automated Learning Applied to Fault Diagnosis, *Trans. I.E.E.E., AES-3*, 6, 941–946, 1967.
4. C. W. Cox and C. J. Renken, Application of Signal-processing Techniques to Signals from Electromagnetic Test Systems, *Materials Evaluation*, 173–181, 1970.
5. P. D. Matthewman and H. Nicholson, Techniques for Load Prediction in the Electricity Supply Industry, *Proc. I.E.E.*, **115**, 10, 1451–1457, 1968.
6. R. M. Baer, Note on Computer Diagnosis, *Trans. I.E.E.E., BME-11*, 1, 8–12, 1964.
7. D. F. Specht, Vector Cardiographic Diagnosis Using the Polynomial Decision Method of Pattern Recognition, *Trans. I.E.E.E., BME-14*, 2, 90–95, 1967.
8. A. S. Ginsberg and F. L. Offsend, Application of Decision Theory to a Medical Diagnosis—Treatment Problem, *Trans. I.E.E.E., SSC-4*, 3, 355–362, 1968.
9. R. S. Ledley, Practical Problems in the Use of Computers in Medical Diagnosis, *Proc. I.E.E.E.*, **57**, 11, 1900–1918, 1969.
10. N. I. Moiserva and V. V. Usov, Some Medical and Mathematical Aspects of Computer Diagnosis, *Proc. I.E.E.E.*, **57**, 11, 1919–1925, 1969.
11. J. Wartak, Practical Approach to Automated Diagnosis, *Trans. I.E.E.E., BME-17*, 1, 37–43, 1970.
12. C. A. Kulikowski, Pattern Recognition Approach to Medical Diagnosis, *Trans. I.E.E.E., SSC-6*, 3, 173–178, 1970.
13. D. J. P. Barker and J. M. Bishop, Computer Analysis of Symptom Patterns as a Method of Screening Patients at Special Risk of Hypothyroidism, *British Journal of Preventive and Social Medicine*, **24**, 193–196, 1970.
14. J. B. Fermiss, A. M. Neville, J. J. Brown, I. G. O'Muircheartaigh, R. Fraser, J. I. S. Robertson, A. W. Kay, T. Symington and A. F. Lever, Hypertension with Aldesterone Excess and Low Plasma-renin: Preoperative Distinction Between Patients with and without Adrenocortical Tumour, *Lancet*, 995–1000, November 14, 1970.
15. R. M. Hirschfe and R. D. Judge, Clinician Versus Computer—A Study of the Application of Bayes' Theorem to Clinical Medicine, *Annals of Internal Medicine*, **72**, 5, 777, 1970.
16. B. G. Batchelor, C. D. Binnie and D. S. L. Lloyd, Discriminant Analysis of Visually Encoded EEG Data, *Proc. Electro-Physiol. Tech. Assn.*, **18**, 3, 150–162, 1971.
17. P. J. Taylor, S. J. Pocock, S. A. Hall and W. E. Waters, Headache and Migraine in Colour Retouchers, *British Journal of Industrial Medicine*, **27**, 364, 1970.
18. G. C. Cheng, R. S. Ledley, D. K. Pollock, A. Rosenfeld (Eds), Pictorial Pattern Recognition (Thompson, Washington, 1968).
19. W. J. Hankley and J. T. Tou, Automatic Fingerprint Interpretation and Classification via Contextual Analysis and Topological Coding, in Pictorial Pattern Recognition (Edited by G. C. Cheng, R. S. Ledley, D. K. Pollock and A. Rosenfeld, Thompson, Washington, 411–456, 1968).
20. A. Ward and G. W. McMaster, Computer Detection of Cancer Cells, *Science Journal*, 32–37, August 1966.
21. R. S. Ledley, Automatic Pattern Recognition for Clinical Medicine, *Proc. I.E.E.E.*, **57**, 11, 2017–2035, 1969.
22. J. W. Butler, Automatic Analysis of Bone Autoradiographs, in Pictorial Pattern Recognition (Edited by G. C. Cheng *et al.*, Thompson, Washington, 1968).

23. N. Funakubo, A. Iwamatsu, T. Suzuki, S. Mori and T. Iijima, ETL OCR Pilot Model, *Bulletin of the Electrotechnical Laboratory*, **34,** 1, 4–15, 1970.
24. P. Saraga, J. A. Weaver, D. J. Woolons, Optical Character Recognition, *Philips Technical Review*, **28,** 197–203, 1967.
25. J. R. Parks, Automatic Recognition of Low-quality Printed Characters using Analogue Techniques, *Radio and Electronic Engineer*, **34,** 2, 67–80, 1967.
26. B. G. Batchelor, Learning Machines for Pattern Recognition, *Ph.D. Thesis*, University of Southampton, 1969.
27. A. J. Lerner, Formal Methods of Diagnosis in Engineering and Medicine, *Computer Journal*, **12,** 1, 29–32, 1969.
28. B. Meltzer, N. H. Searle and R. Brown, Numerical Specification of Biological Form, *Nature*, **216,** 32–36, 1967.
29. R. J. Royston, Applications in Nuclear Physics, in Machine Perception of Patterns and Pictures, *Institute of Physics*, *Conference Series*, No. 13, 171–180, 1972.
30. F. W. Preston and J. C. Davis, Application of Optical Processors to Geological Images, in Machine Perception of Patterns and Pictures, *Institute of Physics*, *Conference Series*, No. 13, London, 223–232, 1972.
31. W. K. Taylor, Optical Texture Recognition, in Machine Perception of Patterns and Pictures, *Institute of Physics*, *Conference Series*, No. 13, 276–284, 1972.
32. T. Kasvand, Histogramming of Nerve Fiber Cross-sections and Water Droplets by Methods of Pattern Recognition, in Pattern Recognition, *I.E.E. Conference Publication*, No. 42, 315–326, 1968.
33. G. S. Sebestyen, Decision-making Processes in Pattern Recognition (Macmillan, New York, 1962).
34. S. R. Hyde, Automatic Speech Recognition, Literature Survey and Discussion, *General Post Office*, *Research Department Report*, No. 45, 1968.
35. W. Bezdel and J. S. Bridle, Speech Recognition using Zero-crossing Measurements and Sequence Information, *Proc. I.E.E.*, **116,** 4, 617–623, 1969.
36. S. H. Lavington, Computer Simulation of a Speech Recognition System, *Proc. I.E.E.*, **116,** 6, 1053–9, 1969.
37. A. F. Newell and C. D. Nabavi, Votem: The Voice Operated Typewriter Employing Morse Code, *Journal of Scientific Instruments*, (*Journal of Physics E*), *Series 2*, **2,** 655–7, 1969.
38. B. Hjorth, EEG Analysis Based on Time-domain Properties, *Electroencephalography and Clinical Neurophysiology*, **29,** 306–310, 1970.
39. Author unknown, Pattern Recognition of EEG to Determine Level of Alertness, *McDonnel-Douglas Astronautics Co. Inc.*, Report No. DAC-60538-52, May, 1969.
40. D. W. Thomas, Vehicle Sound Recognition: A Pilot Study, *Ph.D. Thesis*, Southampton University, 1972.
41. D. W. Thomas and B. R. Wilkins, Determination of Engine Firing Rate from the Acoustic Waveform, *Electronics Letters*, **6,** 7, 193–4, 1970.
42. R. R. Sokal and P. H. A. Sneath, Principles of Numerical Taxonomy (Freeman, San Francisco, 1963).
43. B. R. Wilkins and B. G. Batchelor, Evolution of a Descriptor Set for Pattern Recognition, *Proc. Conference on Technical and Biological Problems of Control*, *Instrument Society of America*, 794–9, 1970.
44. R. M. Needham, Applications of the Theory of Clumps, *Mechanical Translation*, June–October, 1965.

Chapter 8

Looking to the Future

8.1 INTRODUCTION

Our approach is based on a simple conceptual model, which is quite adequate for a large number of problems. However, there are many other problems, which do not fit the existing model and require an entirely new framework. In fact, we can foresee that several new types of model must be developed. We shall discuss some of these in this chapter. However, we begin by discussing ways in which our existing techniques could be improved.

8.2 ABANDONING THE HYPERSPHERE

We can, of course, construct a decision surface of whatever complexity we choose, by using a sufficiently large CC. However, it may be more economical to use a slightly more complex subclassifier, if this leads to a significant reduction in the number required. There is no evidence to suggest that clusters are usually hyperspherical. On the contrary, there is some evidence that they are elongated rather than round.* If this is the case, it would seem more natural to use a hyperellipsoid as the basic "shape" for use in the CC (Figure 8.1).

One type of hyperellipsoid† can be defined using two locates ($\mathbf{B}_1$, $\mathbf{B}_2$) and a single size variable (F). Points on the surface satisfy an equation of the form:

$$\sum_{i=1}^{q} (B_{1,i} - X_i)^2 + \sum_{i=1}^{q} (B_{2,i} - X_i)^2 = F \tag{8.1}$$

(Figure 8.2). By using two locates to define a hyperellipsoid, the storage required is approximately double that needed to define a hypersphere. By adding further locates, we could generate even more complex shapes (Figure 8.3).

Further work is needed in two areas:

(a) To decide whether it would be better to use two distinct hypersphere subclassifiers, rather than one hyperellipsoid subclassifier.

* It has often been noted that there is a strong correlation among patterns belonging to the same class. This suggests that either the class contains more than one cluster, or just one cluster which is elongated.

† This is not the general hyperellipsoid as it has only one elongated/reduced axis. All other axes, normal to this, are of the same length.

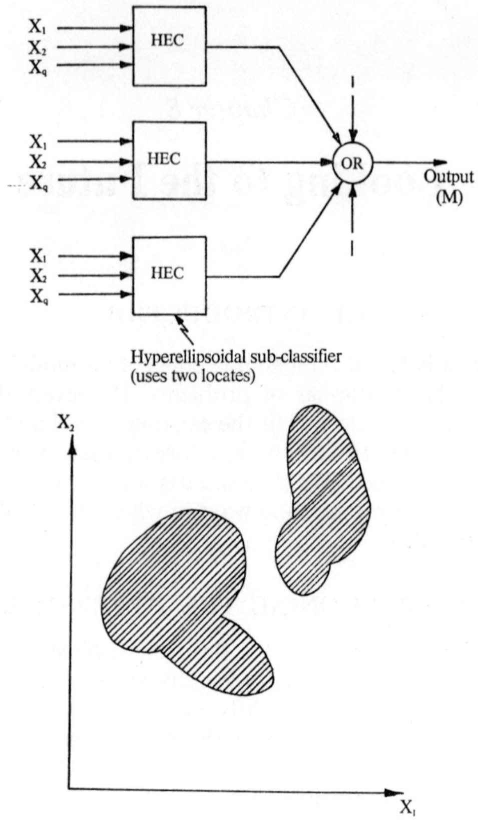

Figure 8.1 Generalised CC, using hyperellipsoidal subclassifiers. (A) Structure. (B) Decision surface.

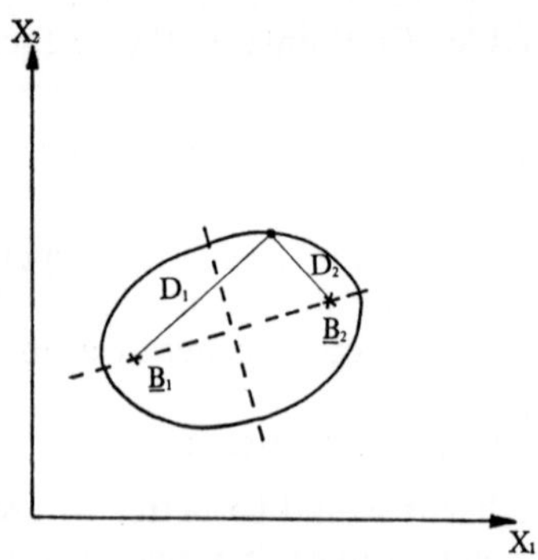

Figure 8.2 An ellipse may be defined by two locates and one size parameter. The ellipse is the locus of a point, moving such that the sum, $D_1 + D_2$, is constant. In a three-dimensional space, a similar definition yields an ellipsoid with one elongated/shortened axis, and rotational symmetry about this axis (rather like a rugby football). The orientation of the ellipsoid may be changed by altering the relative positions of the locates.

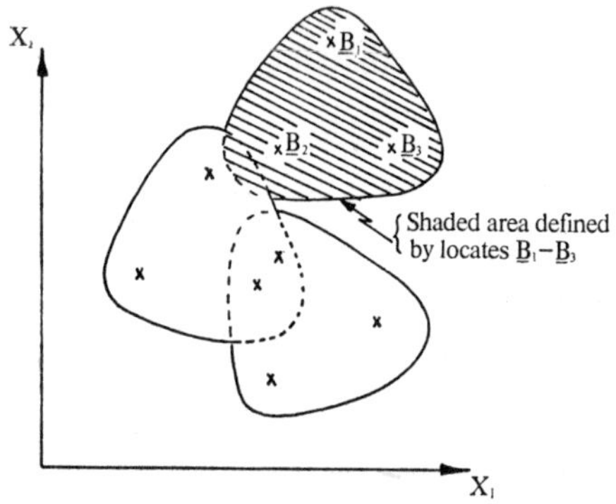

Figure 8.3 Complex shapes can be developed by using more than 2 locates in each subclassifier. In this example, three are used and there are three subclassifiers combined in a CC.

(b) To develop training rules for the CC which uses hyperellipsoid subclassifiers.*

8.3 OBTAINING MORE NEARLY OPTIMAL SOLUTIONS

8.3.1 Using Reinforcement Learning

The major objection to the use of error-correction learning techniques is that they do not always yield an optimal solution. In particular, if the samples contained within the training set have been separated by the classifier, no further modification will ever take place, even though its decision surface may be far from optimal (Figure 8.4). A similar situation

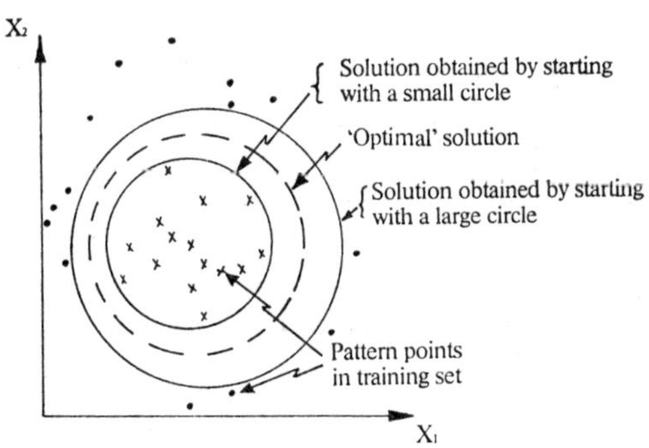

Figure 8.4 Illustrating an objection to the use of error-correction learning.

* Training rules have already been established for a CC which uses hyperellipsoids whose axes of symmetry are always parallel to the coordinate axes of the descriptor space [1, 2].

occurs in the linear classifier. A solution has been suggested by Glucksman [3] who suggested that the classifier be modified if either:

(a) its decision is incorrect (error-correction learning),
(b) its decision is correct, but the pattern point lies close to the decision surface (selective reinforcement learning).

It would be valuable to possess similar rules for the CC.

8.3.2 Eliminating Voids

In section 6.2 we emphasised the need for starting to learn with a small size-parameter, to avoid enclosing large voids (Figure 6.1). In many problems it would be an advantage if the learning rules could shrink the hyperspheres, until they were densely packed with pattern points (Figure 8.5). This would be particularly important in a system

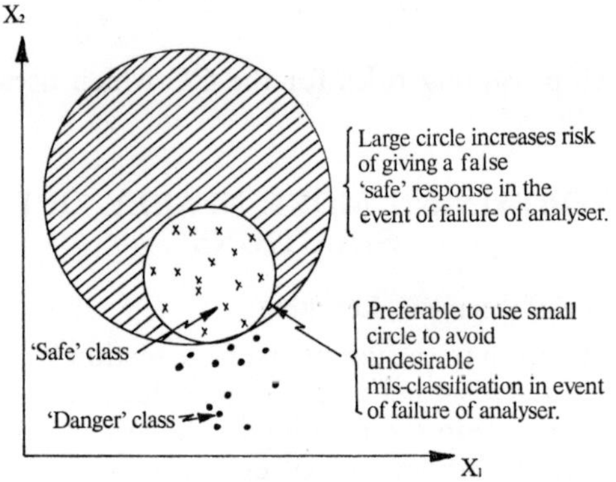

Figure 8.5 Making a classifier "fail-safe" to protect against failures in the analyser.

which must be fail-safe. By reducing the radii of the hyperspheres we can reduce the probability that a fault in the analyser can give a false response of $M = +1$.

An alternative approach to this topic is discussed in section 8.6.

8.4 VARYING THE FREQUENCIES OF ERRORS FROM THE TWO CLASSES

In some problems, it is important to minimise the frequency of errors from one class, even though this will increase the error rate for the other class. We have found this to be a common requirement, particularly in medical applications, where it is often essential to give a "safe" decision, whenever there is any doubt.

The author has just begun a study of this problem in collaboration with Dr. C. D. Binnie and Mr. D. S. L. Lloyd (St. Bartholomew's

Hospital, London). It is too early to report any definite results, although our preliminary work has been encouraging (see Appendix 4). Two ideas are currently being investigated:

(a) Using different values for the iteration parameters (k and l) for the different classes of errors (Method 1).
(b) Selecting samples from one class more frequently than from the other (Method 2).

Both of these methods seem to achieve the required bias, while maintaining a nearly optimal performance. Method 1 appears to be slightly better in this respect.

8.5 ADDING CONFIDENCE

The author has occasionally been asked to supply a measure of the confidence of the decision from a CC.* Often this can be quite crude, merely indicating whether the decision is doubtful or confident. We should consider two aspects of this problem:

(a) Has a pattern like this one ever been seen before? (Figure 8.6). This question is quite distinct from the problem of assigning it to either of the classes. If the present pattern has been seen before, or closely resembles others that have, then we can consider the problem of classifying it. If not, then we should assign

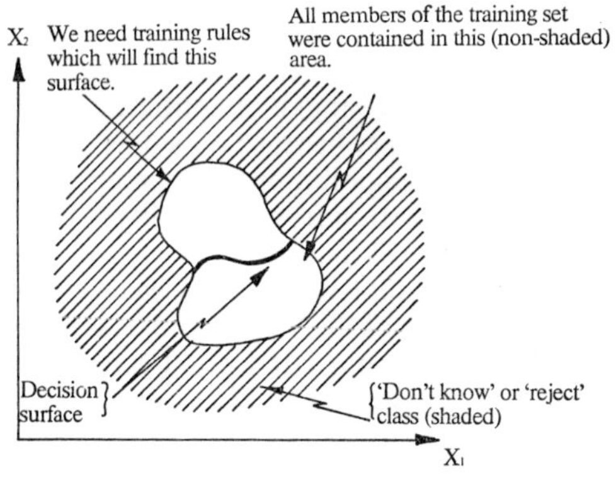

Figure 8.6 So far, we have concentrated on finding boundaries which separate classes. It is also important to be able to recognise patterns which were not encountered in the training set. To do this we need to design surfaces, which enclose "familiar" patterns (of either class), but which do not contain large voids.

* Such a request has usually been made by a doctor, who wishes to retain the right to override the classifier. A doctor would often like to know how confident the decision is, so that in doubtful cases he can consider other factors.

it to a "don't know" class.* The problem of designing a classifier to *recognise* familiar patterns is discussed in the following section. We shall therefore defer further discussion until then.

(b) Assuming that the current pattern is familar, what is its classification, and what confidence can we place in this decision?

The second question will occupy us for the remainder of this section. The confidence of a CC could be defined in a number of ways, each offering various advantages.

The following measure is suggested since it is particularly easy to calculate†

$$\max\left(F_i - \sum_{j=1}^{q} (B_{i,j} - X_j)^2\right) \tag{8.2}$$

This quantity is negative, if **X** lies outside all of the hyperspheres; zero, if it lies on the decision surface; and positive, if it lies inside at least one of them. By finding the distribution of the values assumed by (8.2), it is possible to estimate the probability of mis-classification for new patterns (Figure 8.7).

8.6 LEARNING TO FIND THE BOUNDS OF A CLASS

In some problems it is impossible to obtain representatives of more than one class. For example, consider the problem of recognising pollution in water. First, we must consider what we mean by the term pollution. It can be caused by an excess of acids, alkaline substances, inorganic salts, metals, organic debris, detergents; indeed almost any substance, if present in large enough quantities, constitutes a pollutant. Pollution is difficult to define, because no single chemical, or group of chemicals, alone constitutes pollution. It is better to describe pollution as an excessively large deviation from normality.‡ We wish to construct a pattern classifier which will indicate when pollution is at a high level. The problem facing us is that we have no chance of ever obtaining a truly representative sample of the "pollution" class. How can we design our classifier? We can obtain a fairly good sample of normal (that is non-polluted) water, and we must design our classifier using only this class.

* Sometimes called the "reject" class. The term "unfamiliar" is preferred.

† The quantity

$$F_i - \sum_{j=1}^{q} (B_{i,j} - X_j)^2$$

is calculated by the *i*th subclassifier, immediately before computing its subdecision.

‡ To avoid semantic difficulties we shall define "normality" as the state of the river prior to the dumping of large quantities of industrial and domestic effluent. Of course no definition is entirely satisfactory.

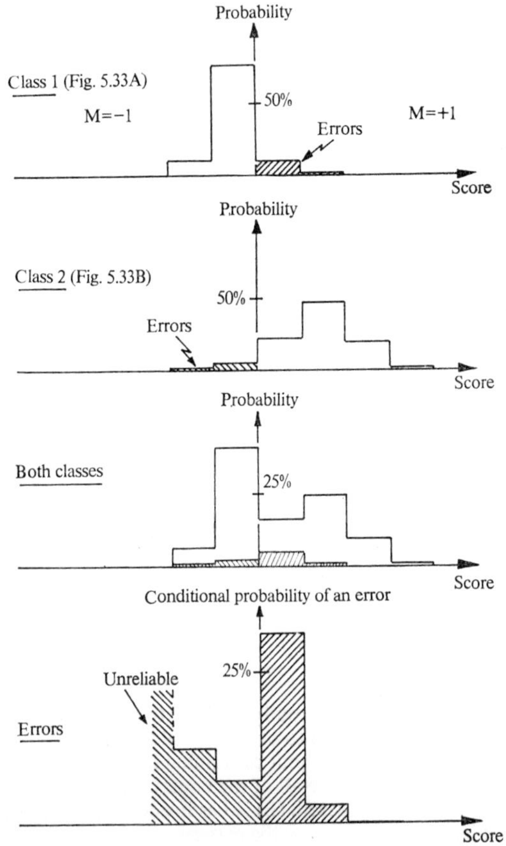

Figure 8.7 A crude method of determining the confidence of the CC decision. The data used here was derived from the classifier, whose decision surface is shown in Figure 5.33H. The score is defined as

$$\max_i \left\{ F_i - \sum_{j=1}^{q} (B_{i,j} - X_j)^2 \right\}$$

All probabilities are estimated by forming histograms of the scores.

Another situation, in medicine, leads to a similar requirement. We need to be sure that our diagnostic classifier is appropriate for the patient under consideration. Suppose we have a classifier to discriminate between two types of peptic ulcer. We require another device to *recognise* peptic ulcers. This second classifier cannot be designed by the familiar rules, because it is impossible to obtain a truly representative sample of the class "not peptic ulcer".

Our present pattern recognition model (Chapter 2) is totally inadequate to handle this situation, so we must develop new learning techniques.

7

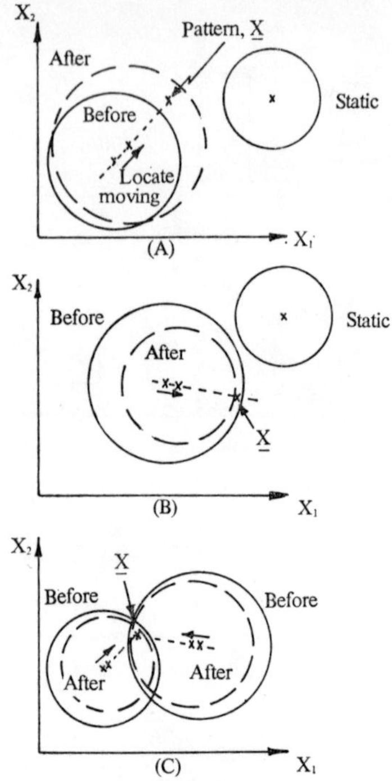

Figure 8.8 Suggested rules for modifying a CC to enclose a given set of data. The ideal solution, that we hope to obtain, would enclose a given proportion of the training set (say 99%) without containing large empty regions, (voids). (A) Pattern point, **X**, lies outside all of the circles. The closest circle is moved towards **X** and its size is increased. (B) **X** lies inside one of the circles which is moved so that its locate approaches the pattern point, while its size is reduced. (C) As (B), except that more than one circle enclosed **X**. All circles enclosing **X** are modified as in (B).

We are currently studying this problem at Southampton. We have developed rules for learning in a CC of fixed size, and are now trying to find methods for automatically varying the size of the classifier (an equivalent process to growing and pruning) (4).

Our learning rules are summarised in Figure 8.8, while Figure 8.9 shows one of our typical results on a 2-dimensional problem.

8.7 MISSING DESCRIPTORS

We have often found that some of the descriptors are missing in some of the pattern vectors, We could simply discard such patterns, although the remaining descriptors may contain most of the information necessary to reach a reliable decision. On the other hand, we could

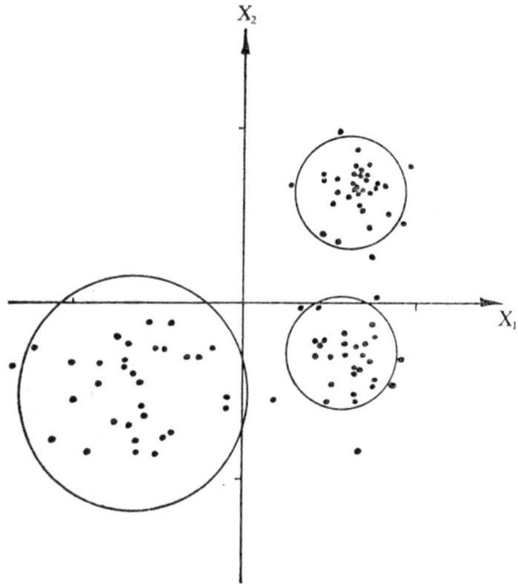

Figure 8.9 Preliminary results obtained using the rules outlined in the previous figure. Data is that given in Figure 5.33B and Table A3.2.

replace missing items by their average values. This approach is clearly too naive to achieve optimum results. We should like to be able to replace the missing descriptors, using the information carried by those remaining. This approach would be rather expensive, since we should require q predictors to replace any missing item in a q-dimensional pattern vector. Even if we were able to do this, problems would arise in trying to replace two or more missing items.*

An alternative solution is suggested in Figure 8.10, where it is assumed that we possess a method for calculating the probability of a pattern vector, **X**, given the values of its components (see Appendix 2). We should then be in a position to scan the missing descriptors, and find the relative probabilities of the pattern classes.

8.8 NON-ORDERED PATTERN VECTORS

We have hitherto assumed that the order of the descriptors within the pattern vector is important. On this basis, the pattern vectors (1, 2, 3), (2, 1, 3) and (3, 1, 2) would correspond to different patterns. Suppose however, that the elements of the vector represent marks gained by students in an examination. The vector describes a *group* of students, not the individuals in it. The vectors (1, 3, 2) and (3, 2, 1) would then represent identical groups.

* It would be difficult to ensure the stability of a system with two non-linear predictors which are connected in a multi-loop feedback arrangement.

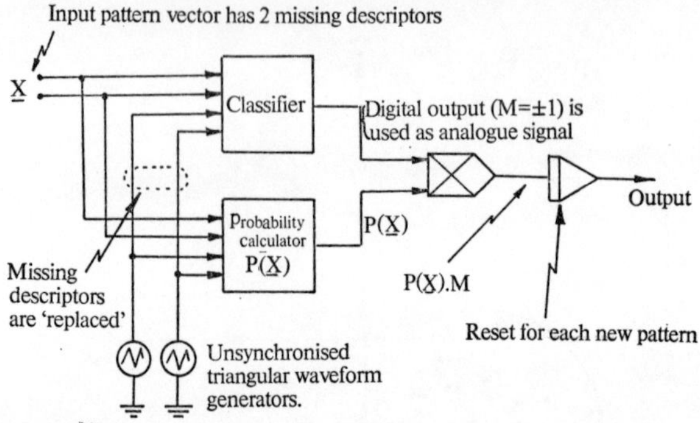

Figure 8.10 Method proposed for making use of non-complete pattern vectors. We attempt to reconstruct the missing descriptors with the triangular waveform generators. We then calculate the probability density function ($P(\mathbf{X})$) for both classes. The more likely classification is found by observing the sign of the output, after the integrator has been in the integrate mode for several cycles of the triangular waveform generators.

In some problems we may be given partially ordered vectors. For example, we might describe individual students by two examination marks, which are always listed in the same order and are contained within round brackets. A description of the group would then typically look like this:

$$((1, 2), (2, 1))$$

This is identical to

$$((2, 1), (1, 2))$$

but different from

$$((1, 2), (1, 2))$$

To illustrate the kind of problem which might give rise to this form of description, we shall return to the problem of recognising languages (previously discussed in section 7.7). Imagine that a sample of an unknown language has been coded by a straightforward substitution of one symbol for another. We do not know the coding, but want to be able to recognise the language as a prelude to decoding. To do this, we could again use letter frequencies for our descriptors. However their order within the pattern vector is meaningless, since they have been shuffled by the random coding.

Another difficulty often occurs. The pattern vector may contain a variable number of descriptors. For example, suppose a man were asked to describe his family. The length of the description of the family is variable, even if the description of each member is not. A very simple description of an individual might contain the following items:

$$(\text{age, weight, height, sex})$$

Then one family may be described as

$$((24, 168, 69, M), (32, 130, 64, F), (29, 145, 66, M))$$

while another is represented by

$$((44, 158, 70, M), (39, 135, 63, F))$$

The second family could equally well be described by reversing the order of the descriptions of the individuals. This type of pattern description could be important in psychiatry and the social sciences.

We have seen that there are problems in which the order of the pattern description is arbitrary, or partially so. It is also frequently found that the length of the description is variable. Often a pattern description will consist of a non-ordered set of ordered vectors. It is the author's firm belief that there is little chance of fitting this type of pattern description into our well-established model. If this is confirmed, we may well see the development of another quite distinct type of pattern recognition.

8.9 CYCLIC DESCRIPTORS

Certain biological measurements have a periodic variation, with a period of 1, 29 or 365 days. Sometimes more than one oscillation is evident. The demand for electricity has three oscillatory components, with periods of 1, 7 and 365 days. A normal demand on a Sunday evening would represent an abnormally light load on Monday morning. When trying to predict the demand for electricity, it is necessary to take the time of day, week and year into account. We begin by discussing the simple case with only one cyclic descriptor.

Suppose that we are going to use time* as one of the elements in our description of a pattern. We could represent time of day on a linear scale, although this destroys its continuity. It would be better to plot the time of day on a circular scale, while the other descriptors are as usual represented on linear scales. This will produce a cylindrical space (Figure 8.11). In such a space, the dissimilarity, between two patterns, may be measured by the length of the shortest path (around the cylinder) which joins their pattern vectors (U in Figure 8.11). However, if these points are close together, this is approximately equal to the length of the chord (V).

A cyclic descriptor may be transformed to form two new, linear descriptors without destroying continuity. Let θ be the cyclic descriptor, and Y_1, Y_2 the new (linear) descriptors, used in lieu of θ. The pattern vector $(\theta, X_1, \ldots, X_q)$ is transformed to the new vector $(Y_1, Y_2, X_1, \ldots, X_q)$ where:

$$\left.\begin{aligned} Y_1 &= \cos(\theta) \\ Y_2 &= \sin(\theta) \end{aligned}\right\} \qquad (8.3)$$

* Other descriptors may be cyclic, for example, geographic location, and the position of a rotating shaft.

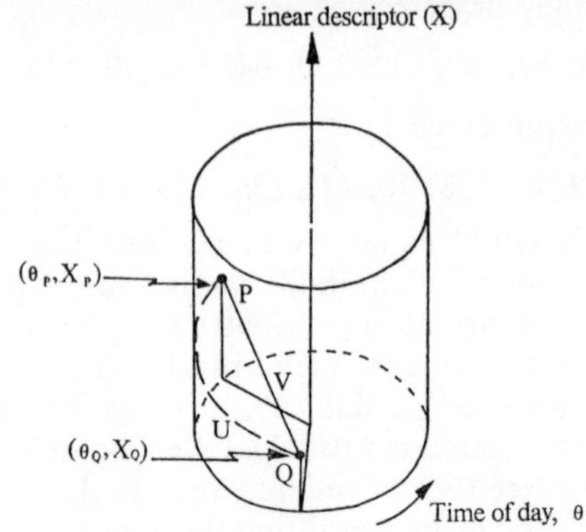

Figure 8.11 A cylindrical measurement space. The "distance" between points *P* and *Q* by path *U* (the shortest route lying on the surface of the cylinder) is approximately equal to that by path *V* (chord joining *P* and *Q*), if *P* and *Q* are close together.

Every element in this new vector is linear, so we should expect to be able to use our conventional learning and classification techniques. However, it must be remembered that the pattern points all lie on a cylinder in this linear space. We do not have any experience of trying to learn with such an atypical distribution;* most of our experimental work has concentrated on learning with cluster-like distributions.

8.10 COLLECTIVE LABELLING OF A GROUP OF PATTERNS

8.10.1 Speech Recognition

Suppose that we are given a short length of magnetic tape, and are told that it contains a recording of a certain spoken word, which we shall denote by *Z*. We are told that the tape also contains recordings of other words, but are not told where *Z* is, nor what a *Z* is. We shall assume that we receive, say, 100 tapes like this, each containing one, or possibly more *Z*s. We are also given 100 other tapes not containing *Z*s. Can we use these 200 tapes to find what a *Z* is? and locate the *Z*s on the first group of tapes?

We shall now code this problem in terms of hyperspaces. A "window" is used in the time-domain, to select a short section from which we derive our descriptors. This window is gradually scanned along the time axis, and at each position a new descriptor vector is calculated. Each

* A similar problem often arises when we are working with descriptors derived from the spectrum of a time series. In many cases, the total power of the original signal is meaningless. To eliminate this, it is a common practice to normalise the descriptors, so that the sum of their squares is constant. This has the effect of forcing all points to lie on a hypersphere.

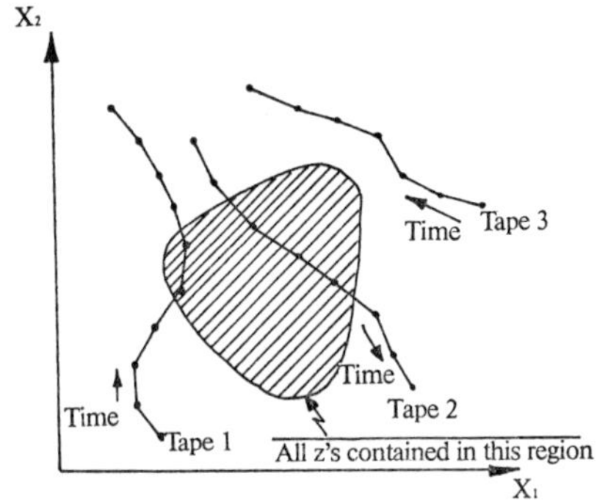

Figure 8.12 The problem of labelling a group of patterns. A tape is represented in the space by a sequence of pattern vectors (solid circles). Tapes 1 and 2 "contain a Z" because some of these pattern vectors lie within the shaded region. Tape 3 does not contain a Z.

tape then becomes coded, as a sequence of points in the hyperspace (Figure 8.12). Now, unknown to us, every Z yields a point, which lies in the shaded area, and could be recognised by our conventional classification techniques. However, we are never told that a given point *is* a Z, but that it *might be.* We cannot use our conventional learning rules, since these would require a more definite statement that the pattern is/is not a Z.

Now we should make it clear that we do not regard this approach as likely to succeed in recognising speech, for the simple reason that a spoken word has no fixed duration. We shall therefore describe two other, more practical, applications in which there is no such objection.

8.10.2 Radar-Target Detection

Consider a narrow-beam radar which is being used to search for a certain target. In practice, the beam and target are both moving, but for our purposes, it is only their relative motion that matters. We shall therefore, fix the beam and allow the target to move. As the target moves across the beam, the power received at the target varies; it is impossible to define precisely where the beam edges are (Figure 8.13). Hence, it is impossible to say that the target is within the beam one moment, and outside it a short while later. We can however, make one of two other statements:

(a) If the target is at a great distance from the aerial, we can confidently state that it is out of the beam.
(b) We can state quite confidently that the target has *at some time* been in the beam, although we cannot say exactly when.*

* By monitoring the incident power at a friendly, co-operative target, we could be more precise about the instant when the target was at the centre of the beam.

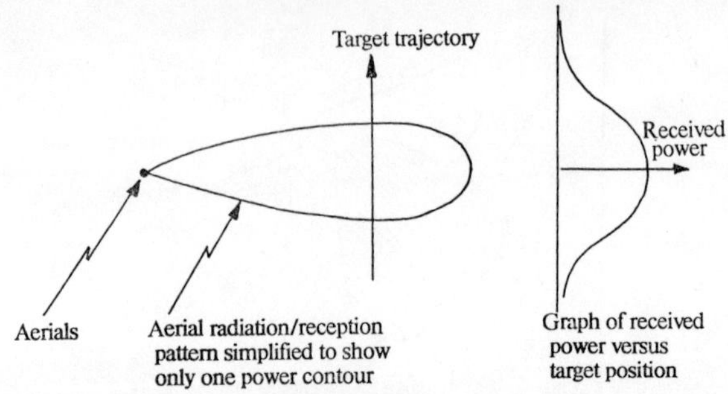

Figure 8.13 Narrow-beam radar, giving rise to the abstract problem outlined in the previous diagram. It is impossible to define exactly when a target enters the beam.

8.10.3 Recognition of Paroxysmal Activity in the EEG

Our third example is taken from electroencephalography. Certain EEG waveforms are considered as being normal, while others are taken as abnormal, and are found chiefly in people with epilepsy. (This is called paroxysmal activity.) Now paroxysmal waveforms may be quite short (about 100–200 ms) and embedded in otherwise normal activity (Figure 8.14). The EEG recording may last for more than an hour, so it is unreasonable to expect the exact limits of every paroxysmal waveform to be defined by visual inspection, of a paper

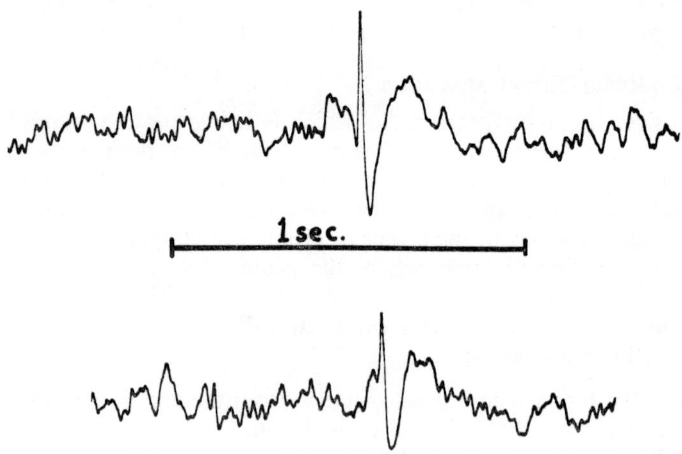

Figure 8.14 Problem in electroencephalography. The event of interest is called a *spike and wave complex*. When does it start and finish? The inability to define these limits leads to the same kind of difficulties as the radar problem, in the previous figure. (Recording supplied by Dr. C. D. Binnie.)

chart. It would be much more satisfactory, if a doctor could label every 5-second* segment to indicate whether or not it contains paroxysmal activity. We may then have a 5-second segment labelled as "paroxysm" even though it contains only one such discharge, which lasts 100 ms.

8.10.4 The Basic Model

All three of these examples may be fitted into the same basic, conceptual model, which we describe below. Figure 8.15 summarises the

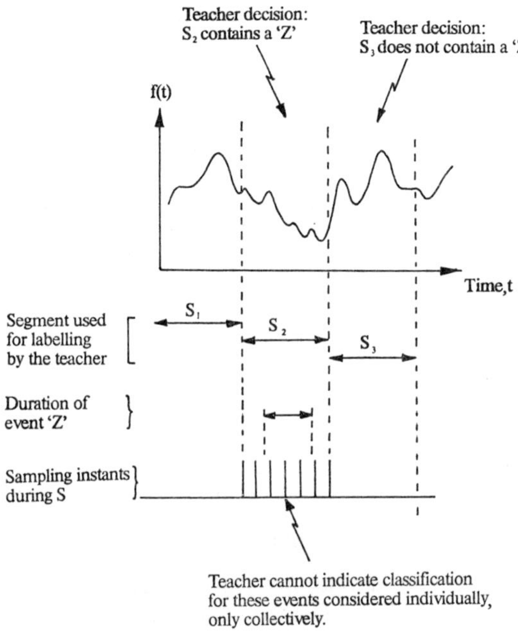

Figure 8.15 The problem of labelling a long sequence containing short items (*Z*s), in the time domain.

problem in the time domain, while the equivalent descriptor space is drawn in Figure 8.16. Patterns are grouped together into sets S_1, $S_2, \ldots$. Each *set* is labelled by the teacher as either "contains a Z" or "contains no Z". *The members of these sets are not labelled individually.* Given a collection of labelled sets such as these, we hope to be able to find a classifier which will enable us to recognise a Z. *This classifier will operate on the individual patterns.* Hence our training data will consist of labelled sets:

$$S_1 = (A_1, A_2, A_3, A_4, A_5) \quad \text{contains a } Z$$
$$S_2 = (B_1, B_2, B_3, B_4, B_5) \quad \text{contains no } Z.$$

* The exact figure is unimportant in our argument.

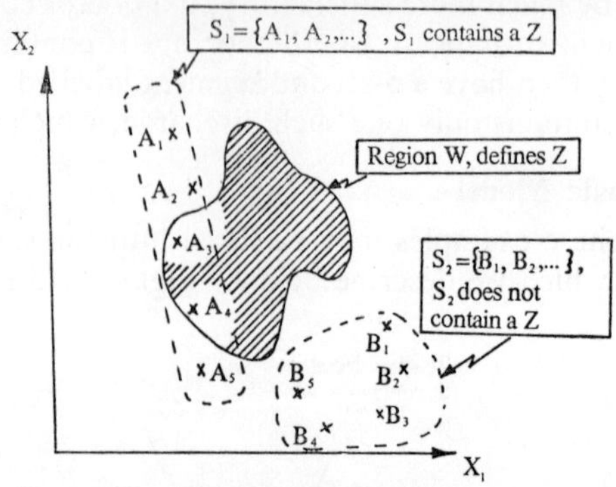

Figure 8.16 The problem of labelling a long sequence containing short items (Zs), in the descriptor space.

From this information, we wish to find the boundary of the region W (Figure 8.16) and approximate this by one of our familiar decision surfaces. After training we shall have a classifier, capable of labelling individual patterns:

$$A_1, Z$$
$$A_2, Z$$
$$A_3, Z$$
$$A_4, Z$$
$$A_5, Z$$
$$B_1, \bar{Z}$$
etc.

If we are successful in developing learning rules to cope with this situation, the classifier will be able to classify individual patterns, whereas its teacher cannot.

8.11 CLASSIFICATION OF SEQUENCES

It has been emphasised by W. I. Card that many diagnostic problems require temporal information [5]. He quotes an example which could, by proper coding, be represented in a form that is quite suitable for our established model.* However, in other instances, complex features must be used, which do not allow a simple recoding to take account of sequence information. Suppose that we wish to use a blood analyser to diagnose a certain disease. The output of this device is a vector, which defines the state of the patient's blood at the instant of sampling. Many

* "Vomiting that is preceded by nausea and is followed by relief as in the vomiting of duodenal stenosis, has an entirely different significance from the unheralded vomiting due to increased intracranial pressure".

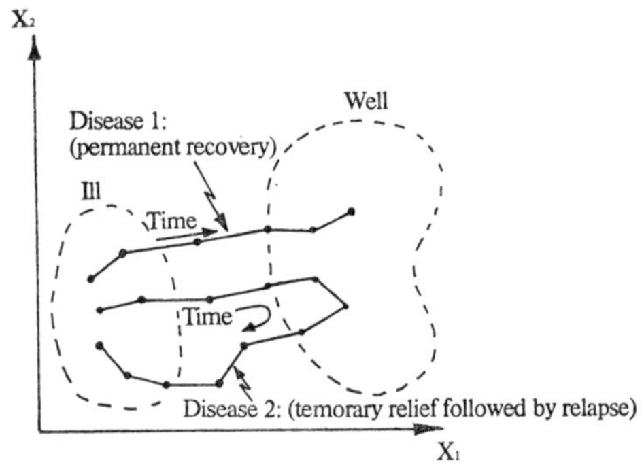

Figure 8.17 Demonstrating the need for temporal information. Two diseases are similar (producing points in the same region of the descriptor space), but respond differently to the treatment, which may be administered to perform a differential diagnosis.

diagnoses are performed by observing the response to treatment.* This response may be regarded as the movement of the blood description-vector in a hyperspace. Hence, we need to know the state of the patient's blood before, and after the treatment. In some instances, it may be helpful to have a more complete record of the trajectory of the vector through the space.

To accommodate the temporal information inherent in some pattern recognition tasks, we must be able to recognise features in the trajectory of a point moving in hyperspace (Figure 8.17). This is, of course, quite a significant change from our existing model, in which the pattern vector is static.

The time variable may be added to our descriptor space. This would give us the opportunity of using the absolute time between events, rather than merely their order of occurrence.

8.12 VAGUE LABELLING OF PATTERNS BY THE TEACHER

In section 8.10 we introduced a vague labelling applied to a set of patterns. In some instances a vague labelling for individual patterns may be available. There are several possibilities including the following:

$$\begin{cases} \text{class } A \\ \text{might be}\dagger \text{ class } A \\ \text{class } \bar{A} \end{cases}$$

* An example is quoted by Baron [6]. "The hypercalcaemia and hypercalciuria of sarcoidosis responds to high doses of cortisone, which does not affect the hypercalcaemia of hyperparathyroidism: this can be used as a diagnostic test."

† Indicates a slight preference for class A.

or

$$\begin{cases}\text{class } \bar{A} \\ \text{might be class } A\end{cases}$$

or

$$\begin{cases}\text{class } A \\ \text{class } A \textit{ or } \text{class } B, \text{ but } \textit{not} \text{ class } C \\ \text{class } B \\ \text{class } B \textit{ or } \text{class } C, \text{ but } \textit{not} \text{ class } A \\ \text{class } C \\ \text{class } C \textit{ or } \text{class } A, \text{ but } \textit{not} \text{ class } B\end{cases}$$

Previously, we have been told to which class a pattern belongs. Here, we may be told that a pattern does not belong to a certain class, but that it may belong to any one of a number of other classes. A pattern may simultaneously belong to more than one class.

Examples are frequently encountered in medicine. A patient may simultaneously exhibit features of sarcoidosis and hyperparathyroidism, but obviously does not belong to the "normal" class.

In problems of this kind we must expect to obtain overlapping classes. So far, we do not know how to train a classifier with vague labelling of the patterns. As always, it is easier to formulate rules than to test them.

REFERENCES

1. B. G. Batchelor and B. R. Wilkins, Error Correcting Algorithm for Adapting a Hyperelliptical Decision Surface for Pattern Recognition, *Electronics Letters*, **4**, 10, 204–6, 1968.
2. B. G. Batchelor, Learning Machines for Pattern Recognition, *Ph.D. Thesis*, University of Southampton, 1969.
3. H. Glucksman, On the Improvement of a Linear Separation by Extending the Adaptive Process with a Stricter Criterion, *Trans. I.E.E.E.*, *EC-15*, 6, 941–4, 1966.
4. P. W. Warren, A Software Approach to One Class Learning, *M.Sc. Thesis*, University of Southampton, 1973.
5. W. I. Card, Diagnostic Process, in Medical Computing (M. E. Abrams, (Ed.), Elsevier, 1970).
6. D. N. Baron, Short Textbook of Chemical Pathology (English Universities Press, London, 1970).

Appendix 1

The Complete Classifier Design Process

A1.1 INTRODUCTION

We have attempted to collect together the ideas discussed in Chapters 5 and 6, to provide a comprehensive set of instructions for designing a CC. These operations are presented in five flow-charts. Some are defined exactly, while others are expressed in less formal terms. We have tried, as far as possible, to define the procedures precisely. However, it is not possible to do this completely, for various reasons:

(a) The method of storing data cannot be defined exactly.
(b) The procedures for counting clusters require human interpretation.
(c) The procedure for deciding between growing and pruning is, at the moment, loosely defined.
(d) The selection of the "best" classifier can only be made by the prospective user.

A1.1.1 Data Storage

The method of storing data should be decided with regard to the facilities available on the computer which is to be used for this work. This affects three procedures:

(a) PSOURCE is the pattern vector selector. Each time this procedure is called, a new pattern vector is available. In addition, the corresponding teacher decision is returned. NP calls of this procedure completely scan the training set. (NP is the number of patterns stored.)
(b) SHUFFLE reorganises the data store by randomly interchanging pattern vectors. Of course, the teacher decisions must be moved in the same way as their corresponding vectors.
(c) A third procedure is used to invert all teacher decisions. This is required, since we do not know which class should be enclosed within the hyperspheres. Hence we shall design classifiers for both possibilities.

A1.1.2 Counting Clusters

The number of clusters is useful in helping us to decide whether we should grow or prune the classifier. This number should be found for both classes. The number of clusters can be estimated using a procedure discussed in Appendix 2. However, this procedure may not always produce a definite result. We suggest therefore that a quick and simple check be made, using the histograms of the descriptors. Both of these techniques for counting clusters require human judgement.

A1.1.3 Deciding Between Growing and Pruning

This decision is also a matter for human judgement. A suggested flow-chart is given in Figure A1.4. However, the safest policy to adopt, if there is any doubt, is "grow now and prune later".

A1.1.4 Choosing the Best Classifier

We emphasise that the prospective user is the only person who can define the "best solution". We can only provide him with a list of possibilities. He will require the following information:

(a) estimated Bayes' error rates (for both classes),
(b) error rates for the CC (for both classes),
(c) size of the classifier.

A1.2 NOTATION USED IN THE FLOW-CHARTS IN THIS APPENDIX

NP1 — Number of patterns in class 1 (*i.e.* that class for which teacher decision = +1).

NP2 — Number of patterns in class 2 (*i.e.* that class for which teacher decision = −1).

NP — Number of patterns, in both classes (NP = NP1 + NP2)(NP, NP1, NP2 refer to the training set.)

Q — Number of descriptors in the pattern vector. The flow-charts allow Q to lie in the range:

$$1 \leq Q \leq 100$$

X[I] — I-th descriptor. I may vary in the range:

$$1 \leq I \leq Q$$

IT — Teacher decision on the pattern with descriptors X[1], . . . , X[Q]. Note that:

$$IT = \pm 1$$

NL — Number of subclassifiers used in the CC. The flow-charts allow NL to have any positive value. In practice, it is unlikely that more than 20 subclassifiers will be required.

B[I, J] — J-th co-ordinate of the I-th locate. Ranges of I and J are:

$$1 \leq I \leq NL$$

$$1 \leq J \leq Q$$

F[I]	I-th size parameter. Range of I is: $$1 \leq I \leq NL$$
M	Decision of the CC. M = ±1.
IDEC[I]	I-th sub-decision. Range of I is: $$1 \leq I \leq NL$$
IWRONG[I]	I-th growing score. Range of I is: $$1 \leq I \leq NL$$
IRS[I]	I-th pruning score. Range of I is: $$1 \leq I \leq NL$$
K	Iteration parameter, determining the magnitude of the increments to the locates. A typical value for K is 0.001.
L	Iteration parameter, determining the magnitude of the increment to the size parameters. A typical value for L is (10Q), where Q is the number of dimensions in the pattern vector.
IER1	Error counter, class 1.
IER2	Error counter, class 2.
MANY	Number of iterations during each learning sequence. A typical value for MANY is 10^4.
IZ	Learning/scoring control parameter. IZ = 0 or 1.
IP	Index. Various uses.
CON	Confidence measure.
MIN	Minimum confidence.
I	Working variable.
J	Working variable.
E	Random variable. Suitable values are ±0.1.
IB1	Counter used to estimate the Bayes' error rate (refers to class 1).
IB2	Counter used to estimate the Bayes' error rate (refers to class 2).
XREF[I]	I-th descriptor of the reference pattern vector. Range of I is $1 \leq I \leq Q$.
ITREF	Teacher decision on the reference pattern. ITREF = ±1.
D	Distance between the reference pattern and another pattern in the training set.
DMIN	Minimum value of D.
II	Working variable.
IS	Teacher decision on the closest neighbour to the reference pattern.

A1.3 NOTES ON FIGURE A1.1

1. It is assumed that the following data has been stored before entering this procedure:
   NP = No. of pattern vectors stored.
   Q = No. of dimensions in the pattern vectors.
   An array of NP × Q elements, co-ordinates of the pattern vectors.
   An array of NP × 1 elements, teacher decisions on the patterns.

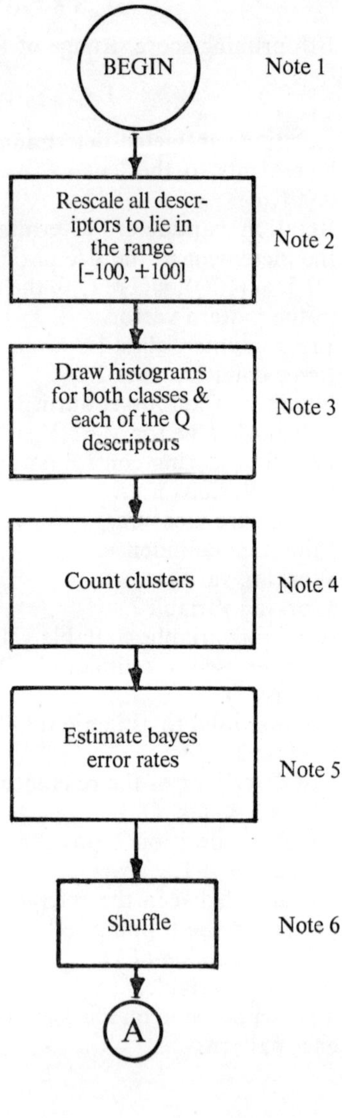

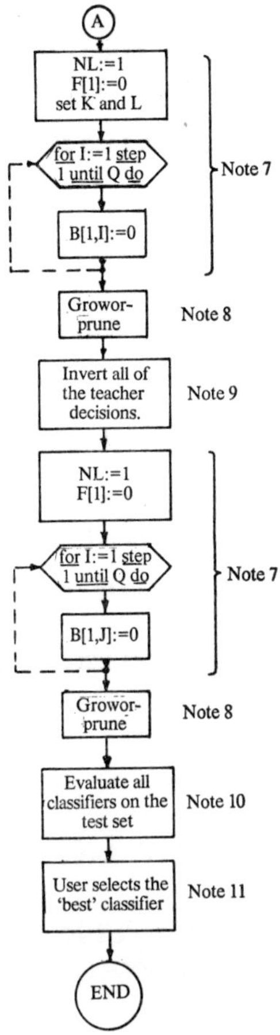

Figure A1.1 Classifier design strategy.

2. Calculate the mean and standard deviation for each descriptor. Normalise so that mean is zero and standard deviation is about 25–30.
3. Used by the operator to obtain a crude estimate of the number of clusters, in both pattern sets. The significance of this number is discussed in the following note.
4. An automated procedure for counting clusters is described in section A2.4.1. However, neither this nor the method based on the visual inspection of histograms (note 3) is totally satisfactory; at best they provide a crude figure. However, this is usually adequate since the number of clusters is only used as a guide for the operator, who must later decide whether to employ the growing or pruning procedure.

5. Described in section 5.4.1. This also provides valuable information to help the operator to decide whether to grow or prune the classifier. Flow chart shown in Figure A1.5.

6. Shuffles the pattern vectors (see section 6.2.6).

7. Initialising the CC to have a single locate, which is placed at the origin. The size parameter is set to zero (see sections 6.2.1, and 6.2.2).

8. Flow-chart for GROWORPRUNE is shown in Figure A1.2.

9. See section 6.2.7.

10. Notice that the test set is used to provide the final evaluation.

11. "Best" classifier can only be chosen by the user, taking the following factors into account:

    (a) estimated Bayes' error rate (note 5),
    (b) CC error rate,
    (c) Number of subclassifiers used by the CC.

A1.4 NOTES ON FIGURE A1.2

1. PSOURCE is the pattern vector source. Each call of this procedure yields:

    (a) a new pattern vector stored in X[1] to X[Q] ($Q \leq 100$),
    (b) the teacher decision corresponding to the same pattern, stored in IT.

    This procedure cycles through the data, and repeats the output sequence after NP calls. This point is particularly important.

2. Flow-chart is shown in Figure A1.3.

3. Error-correction learning.

4. Calculating scores and error rates. At point B, IER1 and IER2 contain the numbers of errors found in class 1 ($IT = +1$) and class 2 ($IT = -1$) respectively.

5. E is a random variable, which need only have two possible states. Suitable values are ± 0.1 (see section 6.2.11).

6. Reorganise the data store by shuffling the patterns.

7. See section 6.2.14 and Figure A1.4.

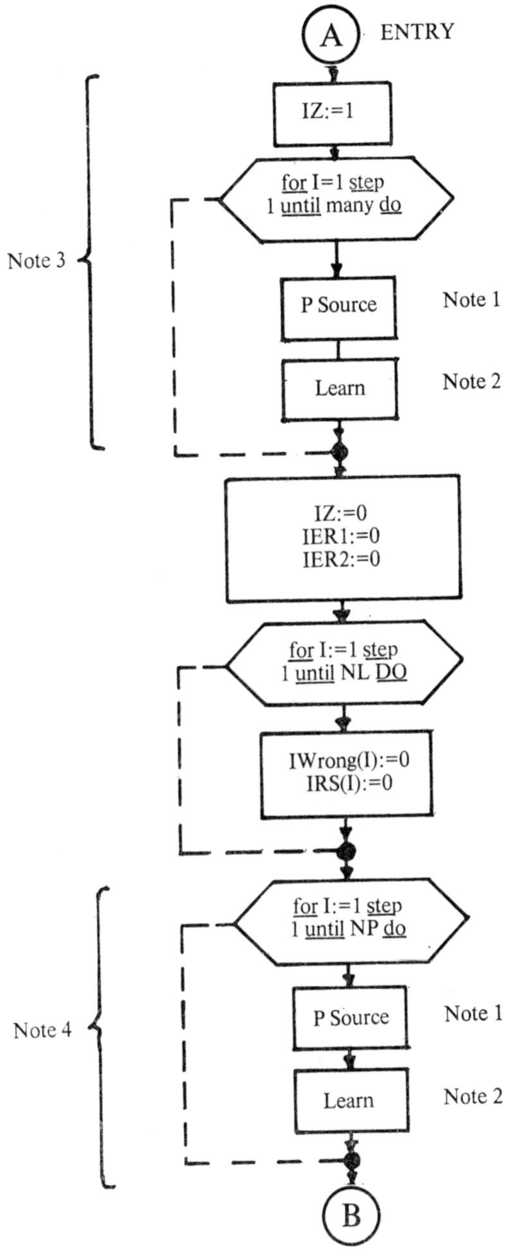
A
ENTRY
IZ:=1
for I=1 step 1 until many do
P Source
Note 1
Learn
Note 2
Note 3
IZ:=0
IER1:=0
IER2:=0
for I:=1 step 1 until NL DO
IWrong(I):=0
IRS(I):=0
for I:=1 step 1 until NP do
P Source
Note 1
Learn
Note 2
Note 4
B

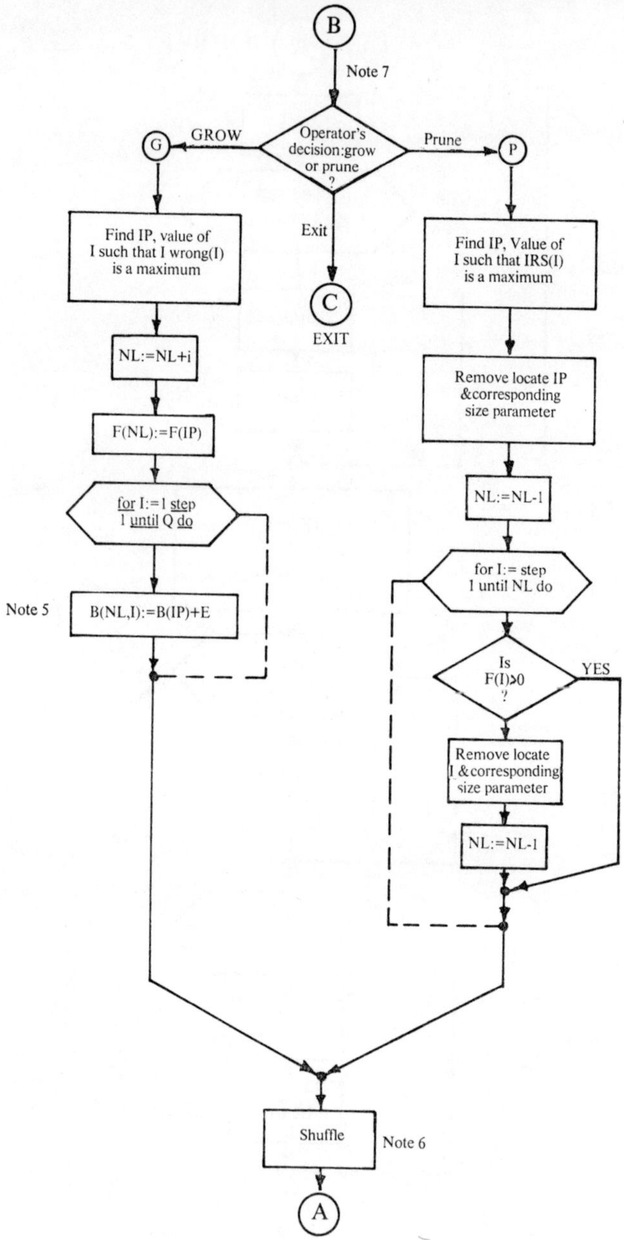

Figure A1.2 Flow-chart for the procedure GROWORPRUNE.

A1.5 NOTES ON FIGURE A1.3

1. Classification phase (between points A and B).

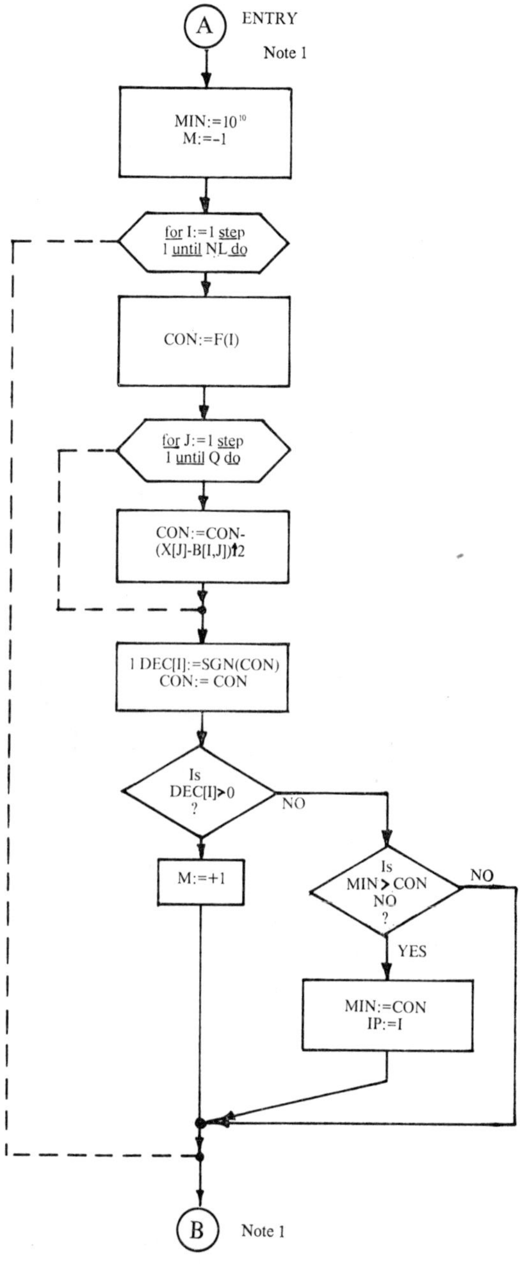

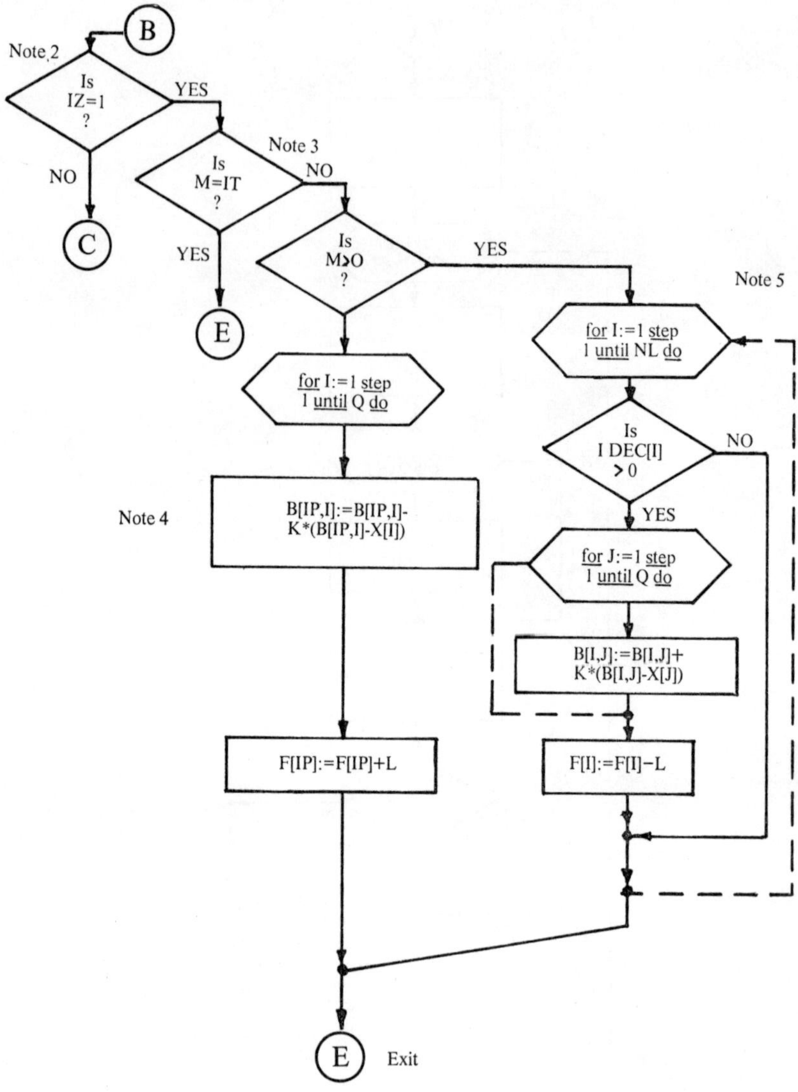
B
Note 2
Is
IZ=1
?
YES
NO
C
Note 3
Is
M=IT
?
NO
YES
E
Is
M>O
?
YES
Note 5
for I:=1 step
1 until NL do
for I:=1 step
1 until Q do
Is
I DEC[I]
>0
NO
YES
Note 4
B[IP,I]:=B[IP,I]-
K*(B[IP,I]-X[I])
for J:=1 step
1 until Q do
B[I,J]:=B[I,J]+
K*(B[I,J]-X[J])
F[IP]:=F[IP]+L
F[I]:=F[I]−L
E
Exit

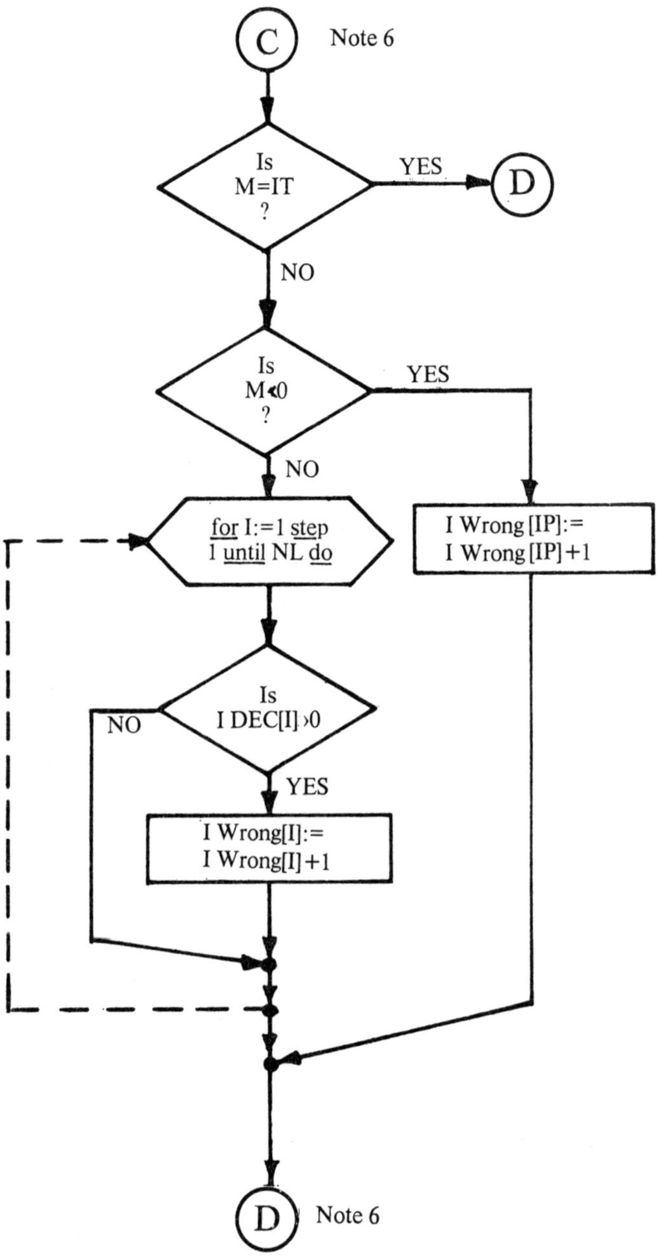
C
Note 6
Is
M=IT
?
YES
D
NO
Is
M<0
?
YES
NO
for I:=1 step
1 until NL do
I Wrong [IP]:=
I Wrong [IP]+1
Is
I DEC[I]>0
NO
YES
I Wrong[I]:=
I Wrong[I]+1
D
Note 6

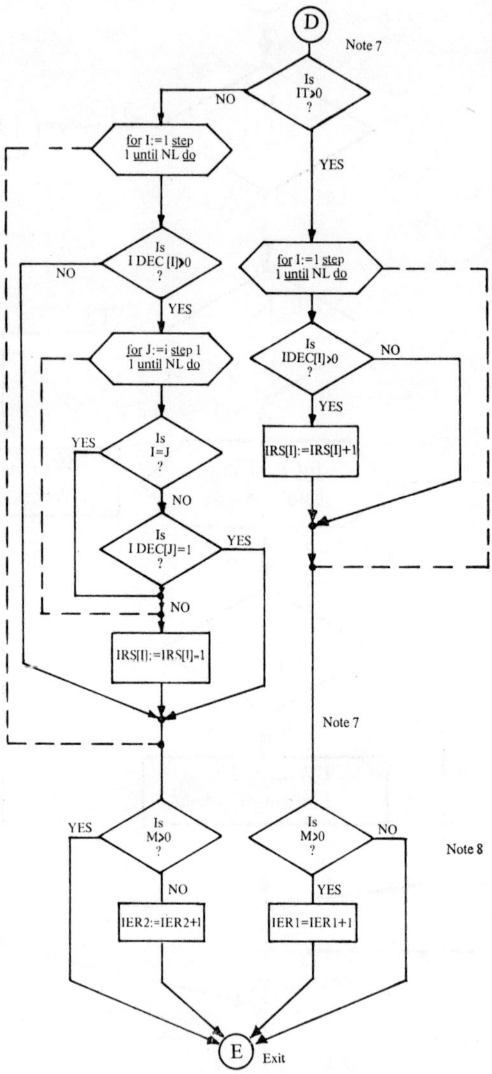

Figure A1.3 Flow-chart for the procedure LEARN.

2. If IZ = 1 the CC is modified, otherwise we enter the scoring procedure.
3. No modification is made if M = IT.
4. M = −1 and IT = +1. Move IP-th locate towards the pattern vector and increase the corresponding size parameter.
5. M = +1 and IT = −1. Move locates for all hyperspheres, containing the input pattern, away from it and reduce their corresponding size parameters.
6. Calculate the growing score, IWRONG[I] (between points C and D).
7. Calculate the pruning scores, IRS[I].
8. Calculate the error rates (using the training set).

A1.6 NOTES ON FIGURE A1.4

1. We need only consider the cycle immediately preceding the present one. When there is only one subclassifier, we always grow.
2. We try to find a "knee" in the graph of error rate versus classifier size. For large classifiers, the error rate will be small. As the size is reduced, we shall normally see the error rate remaining almost constant, until a certain critical size (the "knee") is reached. If the CC is further reduced in size, we shall see a much more rapid variation of the error rate.
3. The following rule of thumb is suggested for this test:

   $$10(Q + 1)NL > MIN(NP1, NP2)$$

   where:

   PRUNE *iff*

   NP1 and NP2 are the numbers of patterns in classes 1 and 2,
   Q is the number of dimensions in the pattern vector,
   NL is the number of subclassifiers (locates) in the CC.
4. The method of estimating the error rate of the Bayes' classifier is described in section 5.4.1 and Figure A1.5. This procedure yields a range of values. Hence, there is some difficulty in defining the optimal policy, when the CC error rate falls inside this range. In this event, it is safest to use the most optimistic (*i.e.* smallest) value for the Bayes' error rate. This has the effect of biasing this decision towards growing. This does not cause significant problems; any surplus subclassifiers can be removed later by pruning.
5. A technique for counting clusters is discussed in Appendix 2. Further comments are to be found in the notes on Figure A1.1.
6. If in doubt, it is safest to grow, since any unnecessary subclassifiers can be removed later by pruning.

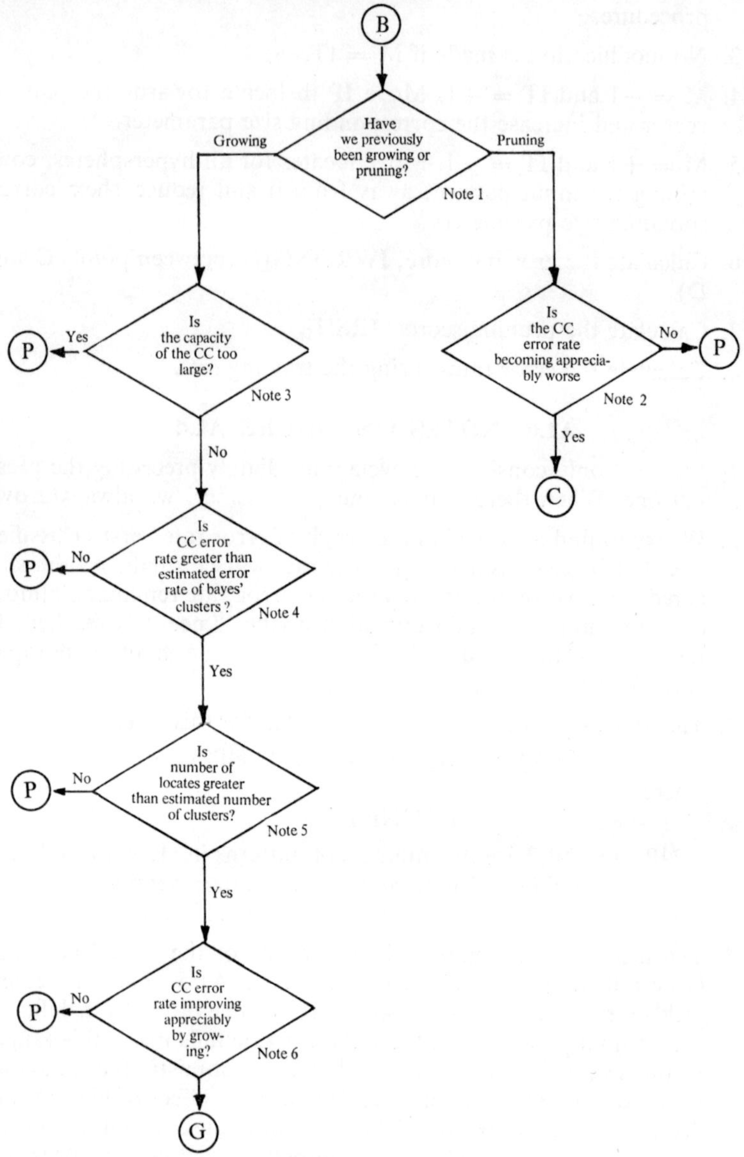

Figure A1.4 Suggested flow-chart for the operator's decision to grow or prune. Key:

G growing
P pruning
C exit
B entry

} refer to Figure A1.2

A1.7 NOTES ON FIGURE A1.5

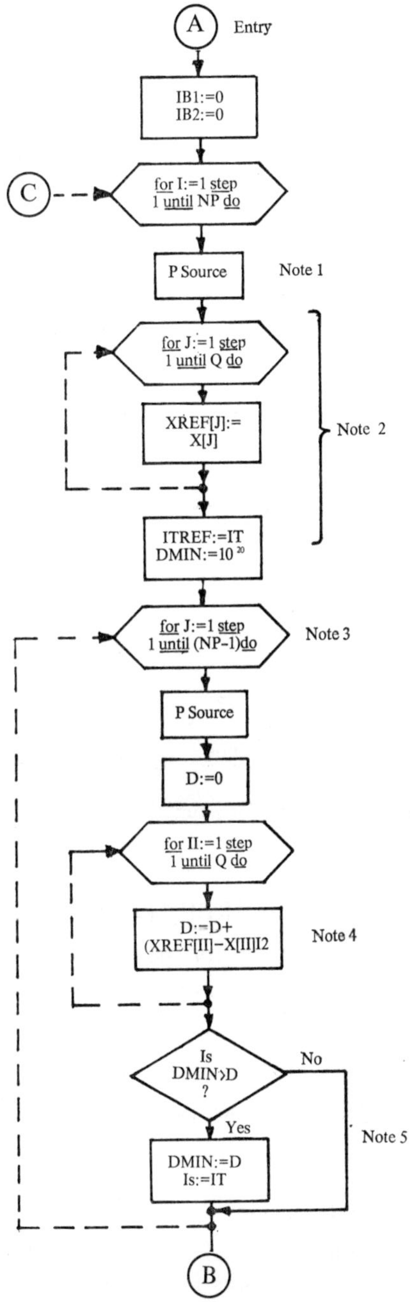

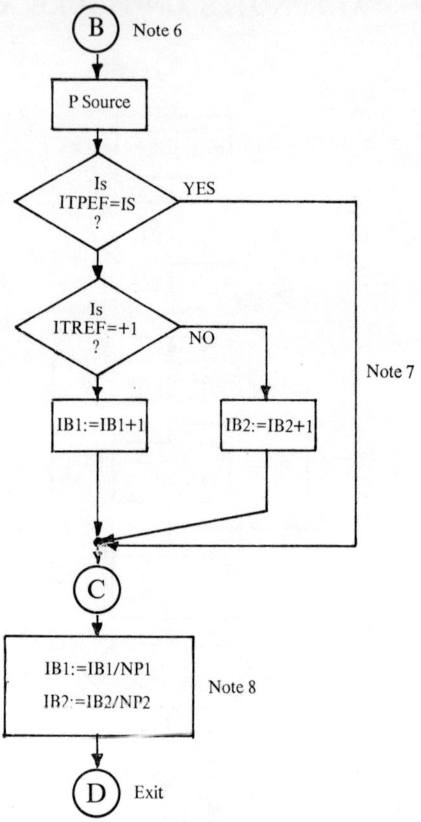

Figure A1.5 Flow-chart for estimating the Bayes' error rate.

1. Select reference pattern.
2. Store reference pattern in XREF[1] to XREF[Q].
3. Scan the training set, except for the reference pattern.
4. Calculate the Euclidean distance between the reference pattern and each of the other patterns in the training set.
5. Find and store the minimum distance.
6. Required to skip the reference pattern, ready for next cycle of I-loop.
7. Increment one of the counters (IB1 or IB2), if the reference pattern and its nearest neighbour belong to different classes.
8. Calculate error rates from IB1 and IB2. (NP1 and NP2 are the numbers of patterns in classes 1 and 2.)

Appendix 2

Miscellaneous Data Analysis Techniques

A2.1 INTRODUCTION

Although not directly relevant to our discussion on pattern classification, the data analysis techniques, described in this appendix are valuable as an aid in understanding the data. They may also be useful in improving the rate of convergence of an adaptive procedure, or starting it nearer to the optimum state. We have already met one of the procedures which should logically be included here, when we discussed the technique for estimating the error rate of the Bayes' classifier (section 5.4.1). This procedure is a good example of the type of analysis technique we shall discuss here. It is valuable in helping us to terminate the learning process at the most suitable time, and can be used to guide us to choose the more appropriate learning process (growing or pruning).

These additional data analysis techniques are ancillary to our main interest in classification. This is reflected in their level of sophistication and cost. They will only be used, in practice, if they offer some distinct advantage *and* are cheap. More refined techniques of data analysis are discussed elsewhere [1].

A2.2 ESTIMATING THE PROBABILITY DENSITY FUNCTIONS

Suppose that we have been given a set, S, containing N pattern vectors, $\mathbf{X}_1, \ldots, \mathbf{X}_N$. (All members of S are drawn from the same class.) Using this data, we hope to estimate the probability density function at a point $\mathbf{X}$ ($\mathbf{X}_i \in S$). The following procedure, although approximate for finite samples, has the merit of being simple to compute and can conveniently be executed at the same time as other data analysis procedures, which we shall describe later.

The procedure for estimating the pdf at a point $\mathbf{X}_i$ is described below:

(1) Select a point $\mathbf{X}_i$ from S.

(2) Find the closest neighbour to $\mathbf{X}_i$ from the set ($S \sim \mathbf{X}_i$). Let the distance between $\mathbf{X}_i$ and its nearest neighbour be D_i.

(3) The pdf at $\mathbf{X}_i$ ($P(\mathbf{X}_i)$) is proportional to

$$\frac{1}{D_i^2} \tag{A2.1}$$

Hence we may write

$$P(\mathbf{X}_i) = \frac{\lambda}{D_i^2} \tag{A2.2}$$

where λ is constant for i in the range

$$1 \leq i \leq N.$$

Strictly speaking, (A2.2) is only valid if N is infinite. When it is finite, we must resort to intuition to justify our use of the procedure.

Figure A2.1 shows the scores derived using this procedure on a simple

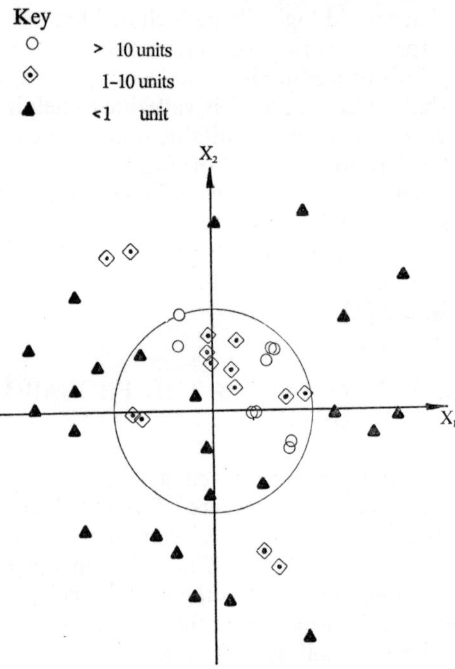

Figure A2.1 A descriptor space showing point values of the estimated pdf. The points shown here have a Gaussian distribution, whose mean vector is at the origin. The pdf estimation technique described in section A2.2 was used. This procedure yields values on an arbitrary scale, but large values are associated with regions where the true pdf is large. The circle, which is one of the contours of the true pdf, has been added to emphasise this point.

set of data. This procedure could be useful in a number of ways; the following are probably the most significant:

(a) To locate and count clusters. An additional facility is required to detect cluster centres from these scores. This will be discussed in section A2.4.
(b) To provide a basis for detecting "unfamiliar" patterns (see section 8.6).

A2.3 ESTIMATING THE CONFIDENCE OF THE BAYES' CLASSIFIER

We shall assume that we have been given two sets of data:

S_1 selected from class C_1, and having NP_1 points,
S_2 selected from class C_2, and having NP_2 points,

where P_1 and P_2 are the probabilities of classes C_1 and C_2 respectively. N is some positive constant. In the procedure described below, all distances are Euclidean and are measured from the reference point $\mathbf{X}_0$.

(1) Select a point $\mathbf{X}_0$. Suppose that $\mathbf{X}_0$ is a member of C_1.

(2) Find that member of S_2 which is closest to $\mathbf{X}_0$. Let the distance between this point and $\mathbf{X}_0$ be D_0.

(3) Count the number of members of S_1 whose distance from $\mathbf{X}_0$ is less than D_0. This number is calculated by including $\mathbf{X}_0$. Let this number be denoted by N_0.

(4) N_0 should be interpreted as follows:

(a) N_0 *greater than 1*. By hypothesis, the correct classification of the point $\mathbf{X}_0$ is C_1. N_0 indicates how likely it is that this decision would be given by the Bayes' classifier. The larger N_0 is, the greater the probability that the Bayes' decision is correct. If the sets S_1 and S_2 are infinite, the expected value of N_0 (say N_0^*) may be related to the probability that the Bayes' classifier is incorrect (P_I), in the following way:

$$P_I = \frac{1}{(1 + N_0^*)} \tag{A2.3}$$

(b) N_0 *equal to 1*. $\mathbf{X}_0$ belongs to the less populous class, in that region of the space.

As always, we must emphasise that we can only collect and analyse finite quantities of data. Hence (A2.3) is strictly speaking inapplicable to the practical situation. However, the inverse relationship between N_0 and the Bayes' confidence does seem intuitively reasonable. The appeal of this procedure is most easily appreciated visually (Figure A2.2).

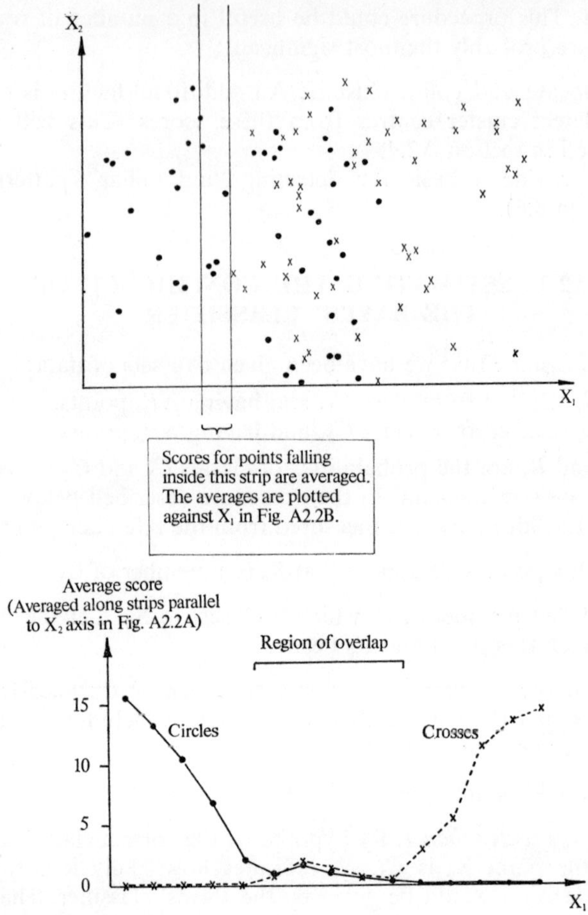

Figure A2.2 Showing results obtained using the procedure for estimating the Bayes' confidence. (A) Descriptor space. (B) Average N_0 scores plotted against X_1.

This procedure may be useful in a number of ways:

(i) To detect and count the number of locates required by a classifier, before we start learning (see following section). This will reduce the learning time.

(ii) To provide a substitute for the teacher confidence. It has been used to vary the iteration parameters during learning. This reduces the learning time, by allowing the adaptation to proceed faster, when the point has been classified with a high level of confidence.

(iii) To eliminate those patterns from the training set whose classifications are unreliable. The importance of this is discussed by Wilkins and Ford [2].

This procedure, and those described in sections (A2.3) and (5.4.1) are closely related,* and may conveniently be calculated in the same program (Figure A2.3).

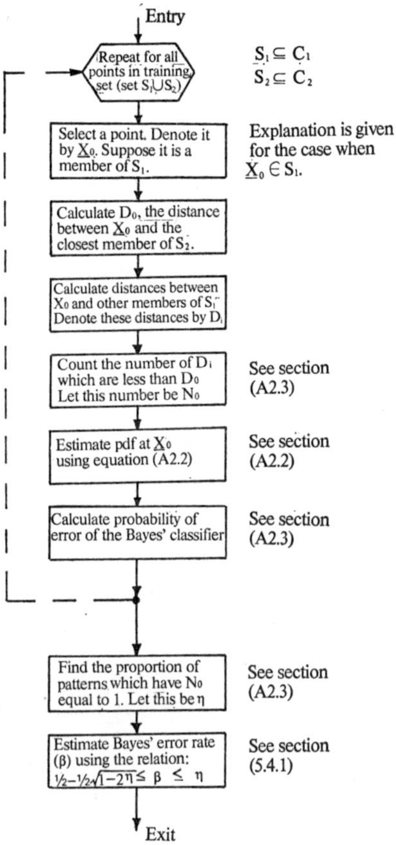

Figure A2.3 Flow-chart of a data analysis procedure which incorporates the techniques described in sections A2.2, A2.3 and 5.4.1.

A2.4 METHODS FOR DETECTING CLUSTERS

Although we already possess powerful techniques for designing a classifier, it would be an advantage if prior to learning, we were able to estimate (*approximately*) how many locates we require, and where we should place them. This could reduce the computational effort involved in designing the classifier, in two ways:

(a) By providing a set of good locates to start learning, we bypass the earlier growing stages.

* For example, the Bayes' error rate is calculated from η (section 5.4.1 and Figure A2.3) which is the proportion of patterns having N_0 equal to one.

(b) By knowing how many clusters there are, we are in a better position to be able to choose whether growing or pruning would be more appropriate.

There are many suggestions for detecting clusters. We shall describe three, to indicate the range of possibilities.

A2.4.1 Maximindist, a Technique for Counting Clusters

This procedure operates on a single class. It was originally described in a slightly simpler form than that discussed below [3, 1]. Maximindist stores a variable number of vectors which for convenience, will be called locates.* Let the locates be denoted by $\mathbf{B}_1, \ldots, \mathbf{B}_p$. Let the data set be $\mathbf{X}_1, \ldots, \mathbf{X}_N$. We shall assume that we have already calculated some suitable scores for each of these patterns.† These scores will be denoted by $S_1, \ldots, S_N$.

In the flow-chart defining maximindist (Figure A2.4) the Euclidean

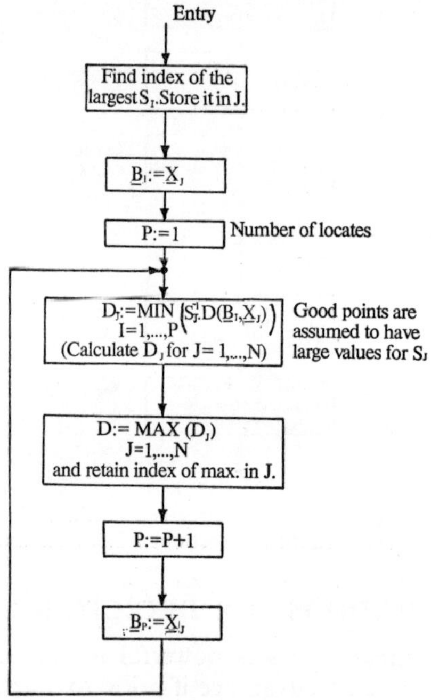

Figure A2.4 Flow-chart of maximindist, a procedure for locating and counting clusters.

* Elsewhere, we have regarded locates as points defining a decision surface. The vectors found by maximindist will eventually be used in this way.

† These could for example, be either an estimate of the pdf, or better still, the estimated Bayes' confidence (see previous section). Any other convenient scoring method could be used instead. Notice that, although maximindist operates upon a single class, the scoring program need not.

distance between the vectors **U** and **V** is denoted by $D(\mathbf{U}, \mathbf{V})$. The original definition of this procedure did not include the scores.

Figure A2.5 shows the performance of maximindist on a simple

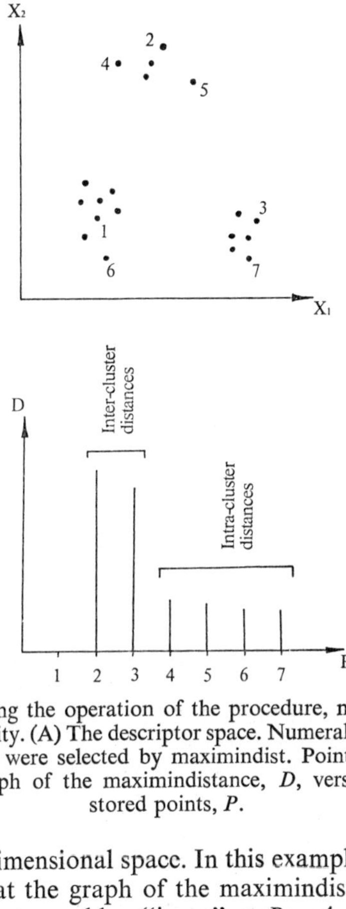

Figure A2.5 Showing the operation of the procedure, maximindist. All S_I scores were set to unity. (A) The descriptor space. Numerals indicate the order in which the points were selected by maximindist. Point 1 was selected at random. (B) A graph of the maximindistance, D, versus the number of stored points, P.

problem, in a two-dimensional space. In this example the S_i scores are all unity. Notice that the graph of the maximindistance, D, versus P (Figure A2.5B) shows a sudden "jump" at $P = 4$. This is the instant at which maximindist is forced to store a vector from a cluster that is already represented by one of the locates. If the clusters are well separated, the position of this "jump" indicates the number of clusters in the space. If the clusters are close together compared to their diameters, there will be no sudden "jump" in the value of D. In Figures A2.6 and A2.7 non-identical scores are used. The same kind of behaviour can be seen, as in our earlier example.

The original procedure, without S_i scores, was susceptible to selecting locates from the periphery of the clusters. The scores force the procedure to choose points which are more typical of the clusters.

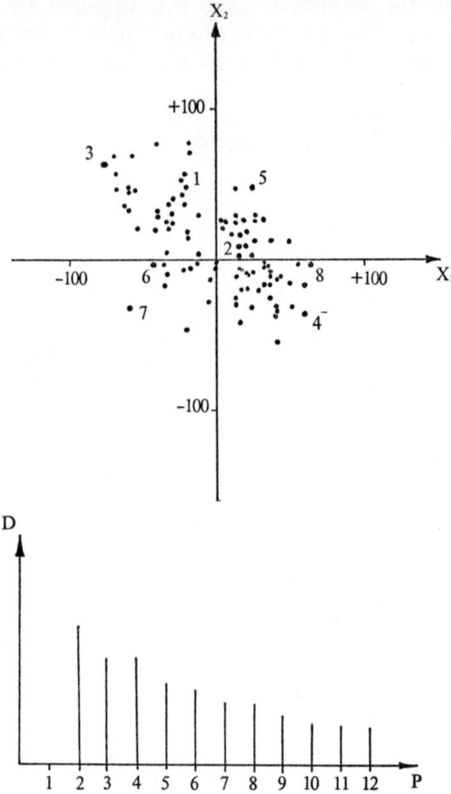

Figure A2.6 Operation of maximindist on the data presented in Table A3.1. The S_I were calculated by the method, outlined in section A2.2, for estimating the probability density function. (A) The descriptor space. Numerals indicate the order of selection of the points. (B) A graph of the maximindistance, *D*, versus the number of stored points, *P*. Notice that there is no sharp decrease in this graph to indicate the number of clusters. In fact there are three overlapping clusters (see Table A3.3).

A2.4.2 Dendrograms; a Method for Displaying Inter-Relationships Between Patterns

Dendrograms have been in use for some years for displaying relationships between biological species [4]. The importance of the dendrogram is that it can conveniently represent pattern distributions, within a high-dimensional space, in a way that a human being can assess visually. A dendrogram resembles a tree used in graph theory. The heights of the bifurcations indicate the "similarity" between patterns higher in the tree.

A dendrogram may be constructed in a number of ways. The method we describe is far from ideal, but has the merit that it is very simple. Let

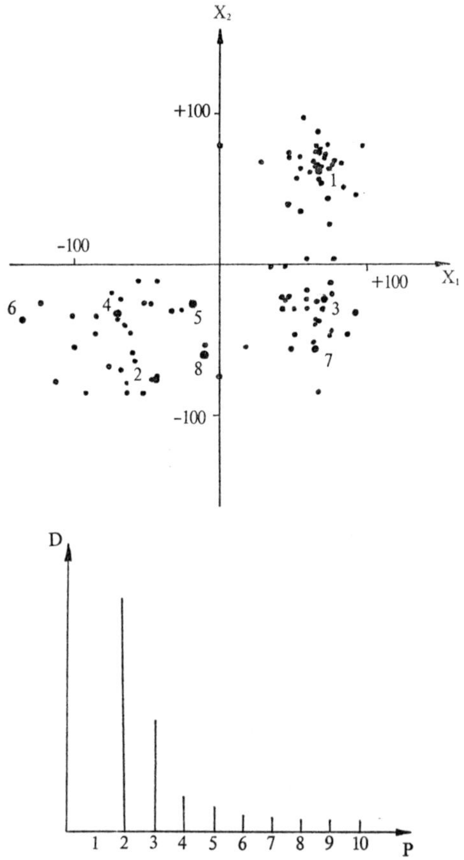

Figure A2.7 Operation of maximindist on the data presented in Table A3.2. The S_I were calculated by the method outlined in section A2.3, for estimating the confidence of the Bayes' classifier. (A) The descriptor space. Numerals indicate the order of selection of the points. (B) A graph of the maximin-distance, D, versus the number of stored points, P. Notice that D is small for P greater than 3, indicating that there are three clusters.

us consider the points $\mathbf{X}_1, \ldots, \mathbf{X}_N$ in a high-dimensional space. A procedure for constructing a dendrogram may be defined as follows (Figure A2.8).

Let the set $(\mathbf{X}_1, \ldots, \mathbf{X}_N)$ be denoted by S_N. From S_N we select that pair of points which are closest together. Let these be $\mathbf{X}_i$ and $\mathbf{X}_j$. We now construct another set S_{N-1}, equal to S_N except that $\mathbf{X}_i$ and $\mathbf{X}_j$ are deleted, and the vector $\frac{1}{2}(\mathbf{X}_i + \mathbf{X}_j)$ is added. This procedure is now repeated on S_{N-1}, to generate a third set, S_{N-2}. The process continues until we have only one point remaining. On the dendrogram, we represent the coalescing of $\mathbf{X}_i$ and $\mathbf{X}_j$ into the single point $\frac{1}{2}(\mathbf{X}_i + \mathbf{X}_j)$ by a

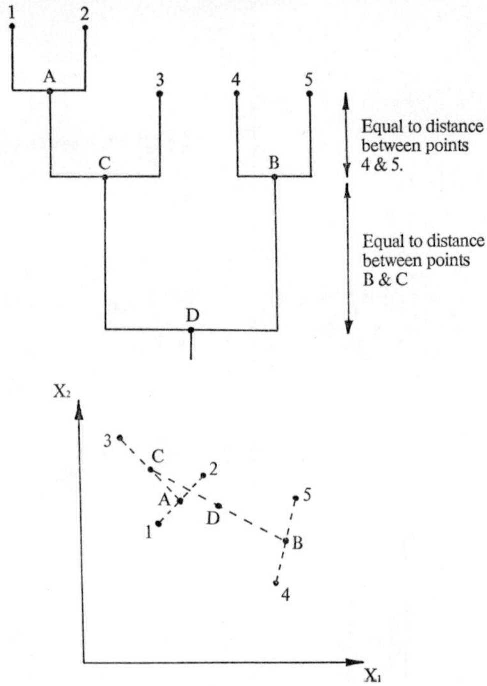

Figure A2.8 A descriptor space showing method of constructing a dendrogram. (A) Descriptor space. The original points are indicated by numbers 1–5. The subsidiary points, generated during the calculation, are indicated by letters A–D. (B) The dendrogram.

bifurcation with two branches. We place $\mathbf{X}_i$ and $\mathbf{X}_j$ at the tips of these branches, while the bifuraction is associated with the point $\frac{1}{2}(\mathbf{X}_i + \mathbf{X}_j)$. Both of these branches are of the same length, which is chosen to be proportional to the distance between $\mathbf{X}_i$ and $\mathbf{X}_j$ in the descriptor space. This simple algorithm places undue emphasis on the lower nodes of the tree. A better algorithm would combine $\mathbf{X}_i$ and $\mathbf{X}_j$ to form a new point

$$\left(\frac{\mathbf{X}_i N_i + \mathbf{X}_j N_j}{N_i + N_j}\right)$$

where N_i is the number of data points that have previously been coalesced to generate $\mathbf{X}_i$. When these algorithms are applied to a class containing several clusters, we find the clustering reflected in the dendrogram (Figure A2.9).

It is difficult to draw dendrograms when N is large, and a computer-based technique is required before dendrograms can be drawn for pattern classes of the size encountered in a practical problem. Until

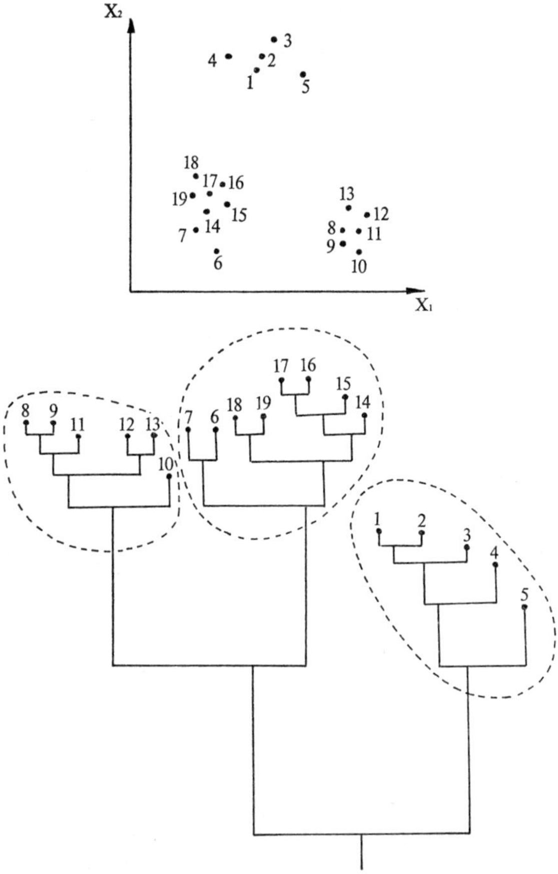

Figure A2.9 A slightly more realistic demonstration of the power of a dendrogram. (A) Descriptor space, containing 3 clusters. (B) The dendrogram. Broken curves indicate the clusters.

such a program can be used to analyse really large sets of data, we cannot properly assess the usefulness of the dendrogram as a means of summarising a pattern class.

A2.4.3 Adaptive Sample Set Construction

This section presents a brief summary of pages 94 to 104 of G. S. Sebestyen's book [5]. Adaptive sample set construction was originally suggested as a technique for finding locates and weights for the Sebestyen classifier (section 3.9).

The procedure requires a number of locates, which we shall denote by $\mathbf{B}_1, \ldots, \mathbf{B}_N$. In addition, we require N counters, represented by $C_1, \ldots, C_N$. In maximindist the locates were fixed, but adaptive

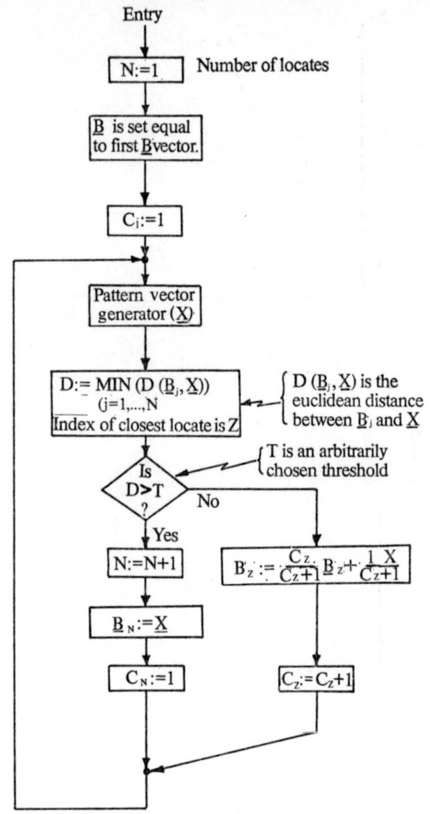

Figure A2.10 Flow-chart for adaptive sample set construction, a procedure for locating clusters. Locates are denoted by $\mathbf{B}_j$. The C_j are counters, recording the number of times that the $\mathbf{B}_j$ have been modified.

sample set construction allows them to move. The procedure is defined by the flowchart in Figure A2.10. Notice the arbitrary threshold, *T*. It is important to realise that the value of *T* significantly alters the performance of the procedure. Let us consider several special cases to emphasise this point.

Case 1: There is one cluster whose radius is less than *T*. The procedure will generate a single locate $\mathbf{B}_1$ which is at the *centre of mass* of the cluster.

Case 2: There is one cluster whose radius is greater than *T*. The procedure will generate several locates, scattered throughout the cluster.

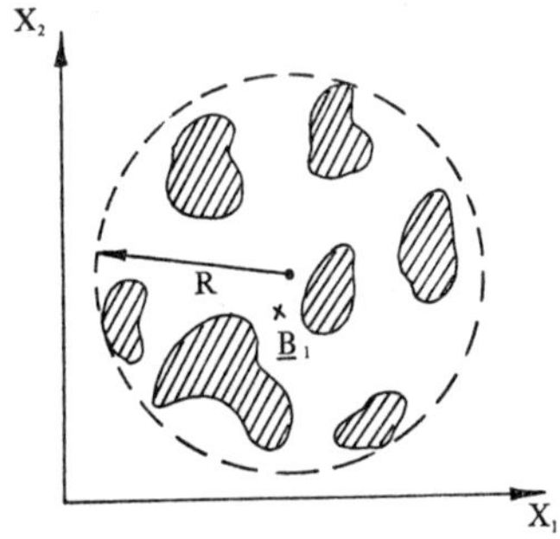

Figure A2.11 A situation in which adaptive sample set construction will fail to detect all of the clusters. The distance R is less than the threshold T in Figure A2.10. In this instance the procedure will produce a single locate, $\mathbf{B}_1$. All points lie inside the shaded areas.

Case 3: There are several clusters which can be contained within a hypersphere of radius $R(R < T$, Figure A2.11). In this situation, the procedure will select a single locate, $\mathbf{B}_1$ (at the centre of mass; all clusters combined).

Case 4: There are a number of clusters whose radii are all less than T, and whose separations are greater than $2T$ (Figure A2.12). In this situation the procedure will correctly detect all of the clusters, placing a locate at the centre of mass of each one.

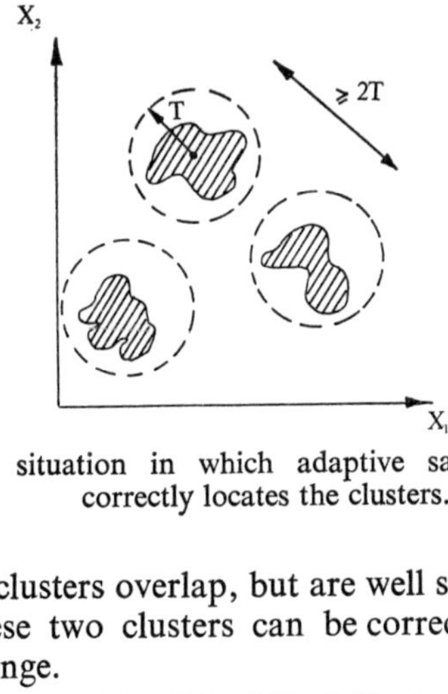

Figure A2.12 A situation in which adaptive sample set construction correctly locates the clusters.

Case 5: Two clusters overlap, but are well separated from all other clusters. These two clusters can be correctly detected, if T lies within the range.

$$\max(R_1, R_2) \leq T \leq (R_1 + R_2) \qquad \text{(A2.4)}$$

where R_1 and R_2 are the radii of the clusters.

This critical dependence on T seriously weakens the procedure, as a technique for cluster detection.* The procedure does not behave satisfactorily when the clusters have ill-defined boundaries (pdfs having long, tapering tails). In this situation, there is a tendency to generate too many locates, whatever value of T is used. A recently published modification of this procedure appears to overcome many of these objections [6].

REFERENCES

1. E. A. Patrick, Fundamentals of Pattern Recognition (Prentice-Hall, Englewood Cliffs, N.J., 1972).
2. B. R. Wilkins and N. L. Ford, The Analysis of Training Sets for Adaptive Pattern Classifiers, in Machine Perception of Patterns and Pictures, *Institute of Physics, Conference Series*, No. 13, 267–275, 1972.
3. B. G. Batchelor and B. R. Wilkins, Method for Location of Clusters of Patterns to Initialise a Learning Machine, *Electronics Letters*, **5**, 20, 481–3, 1969.
4. R. R. Sokal and P. H. A. Sneath, Principles of Numerical Taxonomy (Freeman, San Francisco, 1963).
5. G. S. Sebestyen, Decision-making Processes in Pattern Recognition (Macmillan, New York, 1962).
6. A. N. Mucciardi and E. E. Gose, An Automatic Clustering Algorithm and its Properties in High-dimensional Spaces, *Trans. I.E.E.E., SMC-2*, 2, 247–253, 1972.

* It is not such a serious objection in the design of a Sebestyen classifier, since the "potential function" tends to smooth out the effects of an unnecessarily large number of locates.

Appendix 3

Generation and Listing of Synthetic Data

A3.1 GENERATING DATA

In this section, we shall describe a method for generating type B synthetic data. We shall consider the problem of generating a set containing a number (N_c) of clusters in a q-dimensional space. The following notation will be used:

$\mathbf{G}_i$ $(i = 1, \ldots, N_c)$	Cluster centre (mode).
H_i $(i = 1, \ldots, N_c)$	Spread.*
S_i $(i = 1, \ldots, N_c)$	Selection probability of the ith cluster
U_i $(i = 0, \ldots, 2q)$	Independent random variates, which are uniformly distributed in the range $[-1, +1]$.

Figure A3.1 shows how the synthetic pattern vector $\mathbf{X}$ is generated. This procedure effectively contains four stages:

(a) $(2q + 1)$ random variates: $U_0, \ldots, U_{2q}$ are generated.
(b) A random vector $\mathbf{V}$ is calculated as follows:

$$\mathbf{V} = \{U_{2i}U_{2i-1}\} \qquad (i = 1, \ldots, q) \tag{A3.1}$$

(c) U_0 is used to select the cluster index (j).
(d) $\mathbf{X}$ is calculated as follows:

$$\mathbf{X} = H_j\mathbf{V} + \mathbf{G}_j \tag{A3.2}$$

The distribution of $\mathbf{X}$ is not simply described. There are N_c clusters, centred at the $\mathbf{G}_i$. The pdfs of these clusters may be taken as crude approximations to the multivariate Gaussian distribution (the distribution of $\mathbf{X}$ is discussed in more detail in the following section). In some applications, it might be an advantage to generate a more exact Gaussian distribution. This may be achieved quite conveniently using the Box-Muller method [1, 2]. Using this technique we could generate† $\mathbf{V}$ as follows:

$$\mathbf{V} = \{-2 \log_e (|U_{2i}|)\}^{1/2} \sin (\pi U_{2i-1}\} \tag{A3.3}$$

* The exact meaning of this term will become apparent later.

† This is rather inefficient, since the Box-Muller method allows us to generate two independent Gaussian variates from two input variables.

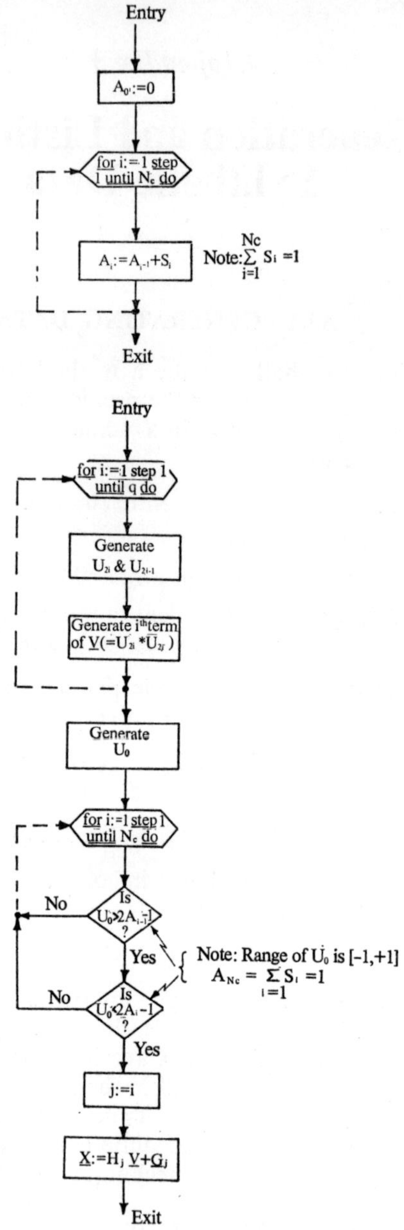

Figure A3.1 Procedure for generating type B synthetic data. The notation is defined in the text (section A3.1). (A) Method of initialising the *A*-array (used in (B)). (B) Method of generating a pattern vector **X**.

It is important to realise that this procedure (Figure A3.1) is only able to generate data from one pattern class. In order to generate data for another class, we simply change the $\mathbf{G}_i$, H_i and S_i and use the same procedure again.

A3.2 CONSTRUCTING THE BAYES' CLASSIFIER

In the previous section, we described a technique for generating synthetic data with precisely known parameters. We may regard a random number generator as a device for *sampling* a population. When viewed in this way, we can see that we have defined a population with a known probability density function. In this section, we shall discuss how we can calculate this pdf from the parameters $\mathbf{G}_i$, H_i and S_i. We can, of course, do this for each of the pattern classes, which will enable us to construct the Bayes' classifier. This is useful in two ways:

(a) To obtain an absolute value for the best attainable error rate.
(b) To obtain a decision surface, which can then be compared with the surface produced by a practical classifier.

We shall illustrate these features in the following section. For the moment, we shall concentrate upon the method of calculating the pdf. There are three preliminary stages in our derivation of an expression for the pdf.

(a) First let us consider two random variates, a and b, which are both uniformly distributed in the range $[\alpha, \beta]$ $(\beta > \alpha > 0)$. Let Z denote the product of a and b. Then the pdf of Z, which will be represented by $P(Z)$, is given by:

$$P(Z) = \begin{cases} 0, & -\infty < Z < \alpha^2 \\ \dfrac{1}{(\beta - \alpha)^2} \log_e (Z/\alpha^2), & \alpha^2 \leq Z < \alpha\beta \\ \dfrac{1}{(\beta - \alpha)^2} \log_e (\beta^2/Z), & \alpha\beta \leq Z < \beta \\ 0, & \beta^2 \leq Z < +\infty \end{cases} \tag{A3.4}$$

(see [3]). Notice that, if α is zero, we obtain an infinite value at $Z = 0$.

(b) Let is now consider a related, but slightly different problem. The variables a and b may now lie in *either* of the two following intervals:

$$[-\beta, -\alpha]$$

and

$$[+\alpha, +\beta]$$

Within these intervals a and b are both uniformly distributed. Their product, which will again be denoted by Z, now has a pdf given by $P'(Z)$, where:

$$P'(Z) = \tfrac{1}{2}P(|Z|), \qquad \{P(Z) \text{ is defined in (A3.4)}\} \tag{A3.5}$$

$P'(Z)$ is plotted in Figure A3.2.

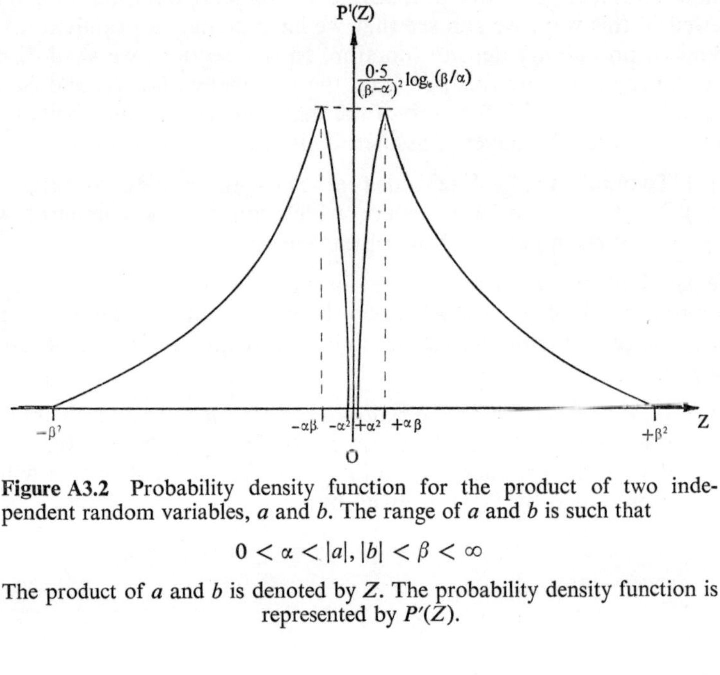

Figure A3.2 Probability density function for the product of two independent random variables, a and b. The range of a and b is such that

$$0 < \alpha < |a|, |b| < \beta < \infty$$

The product of a and b is denoted by Z. The probability density function is represented by $P'(Z)$.

(c) We should like to find the pdf of the product of two variables, which are both uniformly distributed in the range $[-\beta, +\beta]$. $P'(Z)$ is equal to the required pdf, within the two following intervals:

$$\left.\begin{aligned} -\beta^2 \le Z \le -\alpha\beta \\ \text{and} \qquad +\alpha\beta \le Z \le \beta^2 \end{aligned}\right\} \tag{A3.6}$$

By allowing α to be come very small, we may obtain an expression for the required pdf that is valid for all Z, except within a very narrow interval, $[-\alpha\beta, +\alpha\beta]$. The following expression is preferable to (A3.5), since it maintains a large value within the range $[-\alpha\beta, +\alpha\beta]$.

$$P''(Z) = \begin{cases} \dfrac{1}{2} \cdot \dfrac{1}{(\beta - \alpha)^2} \log_e (\beta/\alpha), & |Z| < \alpha\beta \\ \dfrac{1}{2} \cdot \dfrac{1}{(\beta - \alpha)^2} \log_e (\beta^2/Z), & \alpha\beta \leq |Z| < \beta^2 \\ 0, & \beta^2 \leq |Z| \leq +\infty \end{cases} \tag{A3.7}$$

The procedure for calculating the pdf at a point **X**, (given the parameters $\mathbf{G}_i$, H_i and S_i) is shown in Figure A3.3. In order to compute the Bayes' decision for a given point **X**, we apply the following rules:

(i) Calculate pdf1 (the pdf for the first class). We do this by using class 1 parameters in the procedure of Figure A3.3.

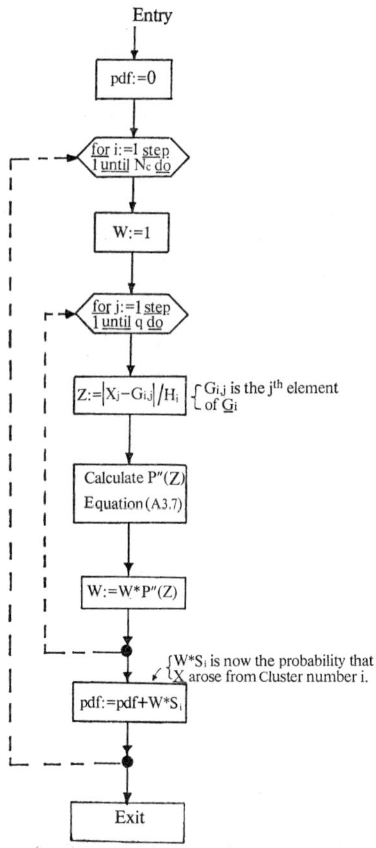

Figure A3.3 Method used to calculate the pdf at a given point **X**.

(ii) Calculate pdf2. As for pdf1, except that the parameters for class 2 are used.

(iii) Calculate L, where:

$$L = \frac{P(C_1)\text{pdf}1}{P(C_2)\text{pdf}2} \tag{A3.8}$$

Here $P(C_1)$ and $P(C_2)$ are the probabilities that $\mathbf{X}$ is a member of class 1 and class 2, respectively.

(iv) Calculate the Bayes' decision as follows:

$$\begin{array}{ll} \text{Class 1,} & L > 0 \\ \text{Don't know,} & L = 0 \\ \text{Class 2,} & L < 0 \end{array} \tag{A3.9}$$

The indeterminate situation, which occurs when both pdf1 and pdf2 are zero, must also be indicated by a "don't know" decision.

A3.3 DATA USED IN THE EXPERIMENTS

Two sets of data, generated by the method described above are listed in Tables A3.1 and A3.2. These are included here, so that the reader can

Table A3.1 Five hundred samples of the data set generated using the parameters given in Table A3.3. The corresponding scattergram is shown in Figure 5.33A. (Class 1)

X_1	X_2	X_1	X_2	X_1	X_2	X_1	X_2	X_1	X_2
34	−8	−22	57	−56	45	20	9	39	−38
−5	−14	−60	44	3	26	15	2	−18	71
13	−42	−13	3	−34	25	36	−3	21	−47
24	42	36	−17	18	−6	−34	10	−34	5
41	−55	32	−3	−19	77	−58	69	−41	77
21	−9	10	17	60	−36	64	−3	13	24
−43	−3	18	26	20	18	−70	69	−30	24
36	−11	−21	48	51	−31	−34	25	50	−15
36	−17	32	−8	13	26	28	−21	34	−5
13	−11	32	−17	−60	47	21	−19	15	9
26	29	−23	32	−20	18	−54	21	15	−6
49	12	0	−6	60	−17	55	−3	−30	30
24	49	40	−32	15	16	4	20	37	13
36	−26	0	−1	43	−11	−69	57	42	19
−20	15	−6	−30	−30	40	−33	21	−14	−2
20	−9	−63	35	26	26	−13	29	24	−9
−60	32	−40	32	41	−13	−21	48	−33	37
17	−20	13	29	9	13	32	−5	26	13
6	25	−22	37	16	−6	13	47	24	2
31	26	13	−30	−40	29	41	−34	−77	63
−69	26	−30	2	32	−36	32	−7	−54	37
−13	25	13	4	−23	40	51	1	20	−14
−28	26	41	−18	43	−22	6	31	40	0
22	−14	−3	43	−25	35	13	−16	41	−36

Table A3.1 (*continued*)

X_1	X_2	X_1	X_2	X_1	X_2	X_1	X_2	X_1	X_2
−56	77	−37	37	−56	29	26	9	13	24
13	−32	18	−5	18	15	−47	25	−1	31
10	3	−37	5	2	11	11	−31	−34	32
39	−9	−28	−10	30	−21	55	−16	41	−5
−13	19	35	−19	22	−11	11	−42	33	−36
−60	29	40	5	−30	−13	4	2	−1	−2
−29	42	−23	−3	26	24	49	32	22	24
33	−16	18	−21	−18	25	24	−5	54	−22
29	17	31	−20	25	31	34	−3	24	15
11	1	−77	21	15	28	43	22	−6	10
−63	43	−23	18	42	11	15	15	−40	44
−13	38	4	24	27	−7	−32	69	−22	12
28	−8	−34	43	−39	29	28	15	4	20
29	13	54	−13	29	24	31	−5	−15	63
13	−5	12	2	26	26	18	25	−6	47
−22	35	64	−6	−45	58	−29	5	13	25
13	−9	7	−9	−15	58	8	15	17	2
−63	32	32	−3	3	11	−29	44	32	−36
36	−9	0	8	1	−38	3	6	−43	2
−40	5	21	−8	38	4	−56	71	−39	63
−33	32	30	−8	2	26	54	−3	32	0
−14	20	2	42	0	47	20	49	−60	29
19	15	39	−16	−1	25	26	−21	−19	43
35	−16	21	−14	32	1	15	15	40	−15
11	−17	22	−28	41	−11	−33	7	32	−36
38	17	13	16	−7	50	−41	35	−47	−2
2	24	13	−9	2	2	43	−3	59	−17
−28	16	54	6	−5	−26	13	0	28	−5
0	13	30	−22	16	15	32	−6	−23	43
−6	16	−9	−6	47	15	−15	35	−29	21
19	25	25	47	−30	58	15	17	18	20
7	−28	−37	51	−24	36	35	−16	−15	36
−22	5	9	17	11	15	29	−2	38	15
18	26	54	0	−45	40	32	10	13	−1
26	10	−5	−6	12	−5	−40	10	13	−9
−34	32	−60	32	8	26	46	1	27	−30
−34	32	9	13	13	15	−1	64	18	−13
19	15	−30	25	38	−21	34	−3	15	28
32	−13	24	15	−34	69	−20	44	35	−38
24	4	−23	36	−45	25	−70	71	14	−18
−13	6	30	−16	1	−26	31	12	−41	51
12	31	24	26	−33	32	25	9	21	−16
−28	32	35	−36	−58	32	43	−8	27	−16
32	−9	25	−5	30	−17	40	−13	41	−42
11	−22	10	48	13	−17	13	15	29	15
−53	37	−9	37	−7	29	2	29	15	13
26	4	30	−19	33	−21	−36	10	24	−6
34	−8	32	−6	54	32	25	29	36	−28
25	43	13	−14	−15	40	29	15	26	12
25	26	−9	50	−23	32	8	8	12	10
−34	64	13	37	11	38	41	10	−21	25
−33	25	21	−8	−33	7	−25	−21	13	19
16	15	42	15	16	37	26	−2	32	24

Table A3.1 (*continued*)

X_1	X_2	X_1	X_2	X_1	X_2	X_1	X_2	X_1	X_2
−63	33	39	−16	24	18	47	54	26	49
54	47	−3	40	34	5	−22	45	−40	32
−13	26	32	−13	46	5	−70	−21	25	15
25	16	−33	51	45	−16	25	10	−87	2
35	0	30	−42	−6	25	45	−14	−14	25
−37	−2	−54	25	32	−9	24	−28	−33	42
−87	18	−23	77	32	1	24	29	21	−6
36	−30	49	20	−36	71	0	2	−13	9
−22	23	20	26	43	−21	−70	32	38	14
−22	18	18	9	21	5	46	−3	54	25
40	−23	45	−31	51	−17	−5	−32	−47	45
55	−8	9	4	43	32	−39	−10	−15	32
−33	29	15	25	15	−1	17	29	60	−28
35	−16	−3	35	−6	13	14	−18	22	−2
13	−5	15	11	41	−23	16	22	50	−8
−63	18	25	−13	49	8	26	−20	−53	35
−22	32	15	49	47	43	−33	43	−20	19
15	−1	39	−8	−28	43	2	44	41	−9
10	2	−60	43	15	2	16	6	43	2
−20	18	42	13	18	15	−49	16	27	−22
10	12	−2	−13	24	−2	12	6	−63	57
32	−3	−69	−21	−70	58	26	43	28	−18
32	−26	−60	7	9	8	29	29	−36	13
40	−3	47	−6	−34	32	1	−5	22	−11
8	0	50	−26	8	49	−7	42	2	43
34	−9	30	−9	−30	5	−29	16	−56	25
32	−14	15	26	13	−3	46	−9	11	29
−23	51	13	38	−30	31	−34	26	−1	4
35	−26	−22	2	16	−3	−31	42	30	−8
32	−8	41	−32	−5	8	32	−30	39	−17
40	−36	54	−15	−69	32	14	−55	17	16
13	19	−19	13	−1	25	−33	33	8	9
24	13	15	24	41	−16	27	−25	−9	45
25	54	43	−38	42	29	54	16	19	−1
32	−5	−9	−20	41	−17	−31	23	38	24
15	18	28	32	−9	51	−70	−3	24	12
16	29	−58	47	15	43	24	2	35	−21

Table A3.2 Five hundred samples of the data set generated using the parameters given in Table A3.4. The corresponding scattergram is shown in Figure 5.33B. (Class 2)

X_1	X_2	X_1	X_2	X_1	X_2	X_1	X_2	X_1	X_2
−101	−55	−19	−26	76	63	77	65	70	−23
−43	−76	63	63	−10	−54	−26	−30	−85	−46
67	56	−47	−26	74	64	42	−22	−63	−79
−74	−19	−33	−31	59	−23	42	−30	92	−33
−10	−60	71	71	−51	−26	−101	−34	66	88
48	−56	79	−30	68	54	59	−17	64	68
−76	−68	87	−47	−67	−38	−43	−75	−52	−85
−43	−11	66	63	56	97	−123	−26	−75	0
66	−24	−51	−47	59	−30	67	74	75	−12

Table A3.2 (*continued*)

X_1	X_2	X_1	X_2	X_1	X_2	X_1	X_2	X_1	X_2
46	−22	−60	−59	73	44	27	68	46	74
−85	−34	−67	−70	79	0	46	40	−112	−68
65	−40	69	79	54	35	−54	−11	69	65
−63	−85	65	−52	65	−56	74	27	65	79
78	4	44	−1	59	4	54	73	−75	−47
93	46	97	79	−70	−37	58	−64	68	−23
−61	−46	34	−1	−136	−37	69	71	76	−20
84	51	65	−40	83	67	69	65	65	74
−70	−34	61	61	67	77	46	71	65	65
43	−21	54	63	65	−39	−50	−30	74	−33
76	−45	51	57	−67	−85	−17	−55	−92	−85
93	65	88	79	68	−35	−54	0	74	50
−67	−50	−85	−44	−35	−54	69	56	−92	−76
77	71	35	−12	−27	−45	−123	−51	−101	−120
68	−26	65	−43	34	−52	−60	−89	44	68
−51	−34	−112	−30	57	73	42	−34	60	−33
74	69	76	−20	66	69	−64	−106	63	−20
−76	−53	−43	−82	−50	−53	−82	−36	74	69
50	76	−70	−2	−54	−70	81	−23	−71	−120
60	−36	74	48	28	−7	35	−38	−136	−70
65	46	66	75	51	69	−112	−36	−70	−68
74	−26	84	84	−61	−97	70	−38	74	64
76	60	−82	−26	73	83	−56	−97	92	−24
72	−30	46	66	74	46	46	4	62	−29
46	−35	59	92	−71	−37	46	50	52	77
−63	−51	53	−39	48	−17	−75	−68	62	−56
−87	−58	44	51	65	−33	73	73	70	−70
63	61	88	66	74	77	−112	−45	68	63
56	71	44	40	66	46	42	−35	74	88
19	−63	−85	−70	93	−63	70	−25	73	40
−50	−37	−17	−26	−56	−37	66	66	52	61
79	−40	−54	−82	88	60	−47	−47	−43	−106
−52	−26	67	79	−54	51	44	84	28	−43
−67	−31	−26	−40	−101	−59	40	−24	74	−45
92	−38	−67	−51	61	74	−112	−16	−85	−37
60	75	66	−1	75	−51	61	74	65	−24
61	63	−48	−50	54	−29	−47	−53	−10	−95
−47	−37	−43	−82	79	88	46	77	−72	−70
−51	−35	68	−24	−101	−2	75	−30	97	76
56	−11	69	−56	−70	−36	46	73	40	73
79	84	−79	−50	−64	−70	40	−33	65	79
−85	−58	69	−23	66	46	69	73	−92	−62
35	44	−85	−65	58	−59	60	−56	54	74
71	88	46	0	60	−7	65	−33	63	74
−87	−34	88	79	65	83	−114	−44	97	−25
49	−1	−61	0	42	−35	−62	−51	73	66
93	−40	59	61	78	−24	68	66	54	68
65	84	−67	−50	74	67	44	65	56	54
75	71	65	−55	66	46	88	−31	−70	−50
88	−17	59	−31	56	40	−87	−37	−70	−89
−87	−70	77	79	92	65	70	−47	−10	−65
27	56	−71	−75	−48	−82	−92	−54	64	60

Table A3.2 (*continued*)

X_1	X_2	X_1	X_2	X_1	X_2	X_1	X_2	X_1	X_2
−65	−51	65	76	74	−24	−101	−50	−105	−85
−19	−51	81	−31	70	−43	69	−24	−53	−2
68	84	−67	−47	−63	−43	40	54	62	−17
74	−30	68	66	81	−30	74	46	−76	−47
60	−34	−47	−79	−33	−51	−63	−35	−10	−75
57	71	65	93	−70	−76	−51	−59	44	−56
−72	−82	60	−52	46	73	65	−33	42	−33
−79	−34	79	84	60	52	65	65	−76	−54
−70	−51	93	−6	52	69	−62	−50	68	−39
71	61	68	79	−67	−31	65	69	−54	−51
65	−1	76	−23	68	44	57	88	74	−30
56	−39	34	0	48	64	−112	−65	64	66
35	56	−43	−55	59	−17	62	−17	−54	−34
65	−33	−53	−60	54	66	70	−33	68	46
44	−33	77	74	61	66	65	73	−53	−82
63	65	73	69	74	40	44	60	104	−20
46	65	63	−33	−47	−51	49	−29	87	−29
84	64	−87	−30	−67	−54	−51	−68	64	63
66	61	63	−52	−54	−85	67	74	−43	−85
−10	−26	−52	−30	69	54	−53	−9	46	68
81	−26	−47	−51	−50	−55	−53	−35	87	−17
74	−33	58	−41	−72	−59	65	88	−71	−19
−74	−16	74	93	76	−40	65	−39	54	74
65	−55	71	27	−71	−45	66	−34	−27	−51
−65	−19	−51	−79	−43	−37	44	66	73	66
−54	−46	−19	6	76	0	−10	−2	92	−26
68	88	75	−22	59	−33	−112	−35	65	69
79	88	40	27	−52	−50	−53	−50	66	−17
79	66	−53	−59	19	−59	68	83	64	40
−101	−36	79	68	−114	−36	62	−63	48	−39
65	73	57	54	65	−24	19	−45	74	4
66	84	−54	−30	54	76	69	52	−17	−43
63	0	49	−47	−112	−59	75	−41	−60	−34
77	61	34	−33	71	97	75	−45	51	92
54	88	79	79	−10	−36	73	59	79	51
84	65	27	50	54	−22	88	74	59	−45
−26	−26	60	−70	81	−33	66	64	−67	−36
−67	−76	−64	−30	93	54	68	66	92	−31
−101	−54	48	64	−56	−79	−70	−82	−67	−58
74	59	−67	−40	83	74	40	−45	−53	−95
−17	−62	−51	−106	49	−31	75	−33	65	0
−19	−9	66	74	−123	−45	−67	−76	73	74
70	−24	97	−23	74	73	104	−59	46	74
−67	−70	66	−43	−26	−70	−123	−46	−27	−53
−63	−51	54	54	60	60	−75	−55	−96	−95
74	−52	−72	−65	40	−12	66	79	35	−80
34	−11	−51	−9	61	64	66	56	46	48
−76	−62	−47	−30	63	−50	73	79	−19	−85
65	−33	35	77	56	71	−79	−75	83	56
−79	0	88	−25	97	−24	68	73	64	73
68	−56	70	−47	46	−39	65	68	−67	−34
46	79	79	73	−74	−30	74	−17	63	−11
−136	−26	65	−38	−93	−68	68	56	75	−59

easily repeat some of our experiments for himself. The scattergrams for these data sets are shown in Figure 5.33A and B, and their statistics in Tables A3.3 and A3.4. The corresponding Bayes' decision surface is sketched in Figure A3.4, where it is compared to two CC decision surfaces, obtained by growing. In Table A3.5 we compare the error rates produced by these two classifiers, and the Bayes' classifier.

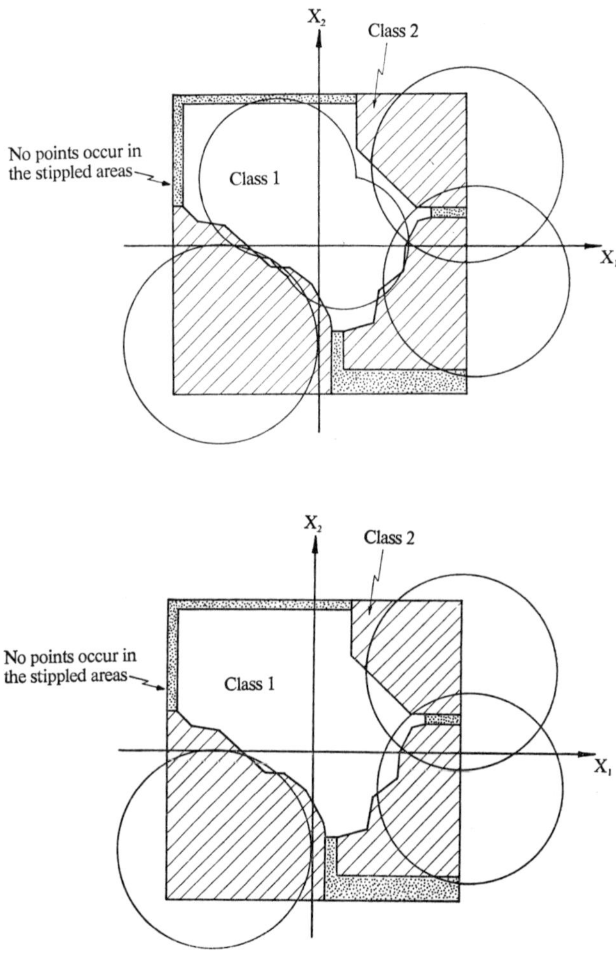

Figure A3.4 Comparison of the decision surfaces of the Bayes' and compound classifiers. The data was generated by the method described in Figure A3.1, with the parameters given in Tables A3.3 and A3.4. The Bayes' decision surface was traced from a computer print-out (100 × 100 matrix). (A) The CC was developed by growing. Class 1 (see Tables A3.1 and A3.3) is enclosed within the circles. (B) The CC was developed by growing *cf.* Figure 5.33). Class 2 (see Tables A3.2 and A3.4) is enclosed within the circles.

Table A3.3 Parameters of the data listed in Table A3.1 (Class 1)

	Cluster		
	1	2	3
Mode (G_i) Along X_1 axis	16.7	−33.3	33.3
Mode (G_i) Along X_2 axis	16.7	33.3	−16.7
Spread: (βH_i)	50.0	60.0	42.5
Selection probability: (S_i)	0.40	0.30	0.30

Table A3.4 Parameters of the data listed in Table A3.2 (Class 2)

	Cluster		
	1	2	3
Mode (G_i) Along X_1 axis	−66.7	66.7	66.7
Mode (G_i) Along X_2 axis	−50.0	−33.3	66.7
Spread: (βH_i)	75.0	50.0	42.5
Selection probability: (S_i)	0.35	0.30	0.35

Table A3.5 Comparing the (percentage) error rates of the Bayes' and compound classifiers. These error rates were estimated using 500 patterns from each class. (The standard error is typically 1.8%.)

	Class 1 (Tables A3.1, A3.3, Figure 5.33A)	Class 2 (Tables A3.2, A3.4, Figure 5.33B)
Bayes' classifier[a]	5.2	6.0
Bayes' classifier[b]	3.5–7.0	4.5–9.0
CC1[c]	5.6	5.6
CC2[d]	5.6	9.0

[a] Error rate of the Bayes' classifier, calculated by the technique described in section A3.2 and Figure A3.3. This is the *true* Bayes' error rate whereas the second entry in the table refers to an *estimated value.*

[b] Estimated value of the Bayes' error rate obtained using the procedure described in section 5.4.1. A range of values is given by this procedure.

[c] CC1 is a CC which encloses class 1 inside the circles. (see Figure A3.4A).

[d] CC2 is a CC which encloses class 2 inside the circles. (See Figure A3.4B.)

REFERENCES

1. G. E. P. Box and M. E. Muller, A Note on the Generation of Random Normal Deviates, *Annals of Mathematical Statistics*, **29**, 610–611, 1958.
2. T. G. Newman and P. L. Odell, The Generation of Random Variates (Griffin, London, 1971).
3. G. G. Roussas, A First Course in Mathematical Statistics (Addison-Wesley, Reading, U.S.A., 1973).

Appendix 4

Addendum: Varying the Proportions of Errors from the Two Classes

A4.1 INTRODUCTION

In section 8.4, we pointed out that we do not possess a method for biasing the classifier, so that we can vary the proportion of errors from the two classes. Two methods were suggested. Between completing the original manuscript and the moment of going to press, we were able to obtain some preliminary results. These are described below.

So far only two classes have been examined in depth. Similar results have been obtained using multi-dimensional data. However, we must emphasise that this statement is made without sufficient evidence for us to have great confidence in it.

A4.2 METHOD 1

Consider Figure 5.31. To implement method 1, we use different values of the iteration parameters, k and l, depending upon whether we employ MODIFY1 or MODIFY2. Let R, k_0 and l_0 be positive parameters, which we can choose at will. Then we let:

$$\left.\begin{aligned} k &= \begin{cases} Rk_0, & \text{in MODIFY1} \\ R^{-1}k_0, & \text{in MODIFY2} \end{cases} \\ l &= \begin{cases} Rl_0, & \text{in MODIFY1} \\ R^{-1}l_0, & \text{in MODIFY2} \end{cases} \end{aligned}\right\} \tag{A4.1}$$

The parameters values are discussed in section A4.4.

A4.3 METHOD 2

Consider Figure 6.4. This shows the flow-chart of a random data selector. The class selection probability can be adjusted by varying the threshold within the conditional statement. Let V be the threshold $(0 < V < 1)$. Then we select a pattern from that class with $T = +1$, if

$Z > V$. Hence the patterns are selected from the two classes in the ratio:

$$\frac{(\text{No. with } T = +1)}{(\text{No. with } T = -1)} = \frac{1 - V}{V} \tag{A4.2}$$

A4.4 EXPERIMENTAL DETAILS

Data: 500 samples of each of the classes listed in Tables A3.1 and A3.2. The same data was used for learning and estimating the error rates. Although this is not good practice, we have found little discrepancy, in other experiments, between the error rates evaluated using the training set and test set. The same data was used to estimate the Bayes error rates, by the method discussed in section A3.2.

Parameters:

$$\left.\begin{aligned} k_0 &= 0.005 \\ l_0 &= 10.0 \end{aligned}\right\} \text{Method 1 experiments}$$

$$\left.\begin{aligned} k &= 0.005 \\ l &= 10.0 \end{aligned}\right\} \text{Method 2 experiments.}$$

Initial CC:

One circle, centred at (0, 0), radius = 31.6.

Learning:

10^4 iterations per growing cycle.

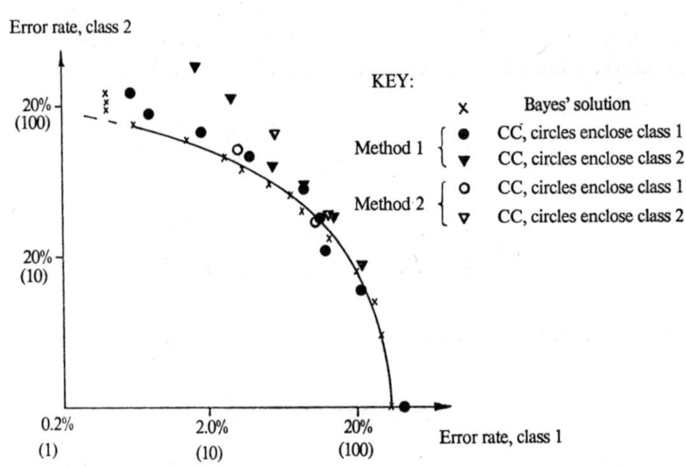

Figure A4.1 Showing the effects of biasing the Bayes' and compound classifiers to vary the proportions of errors from the two pattern classes. The numbers in brackets indicate the numbers of errors found from a set containing 500 patterns.

Final CC: Arbitrarily taken at the end of 3rd growing cycle. (Since there are three clusters in each class, this was felt to be a fairly strict criterion.)

A4.5 RESULTS AND CONCLUSIONS

The results are summarised in Figure A4.1. It is apparent that method 1 is capable of varying the error rates so that we retain nearly optimal performance. The two solid triangles at the top of this figure are felt to be spurious results, due to the small data sets. However, we require many more results to verify or disprove this.

There is no strong indication that method 2 is any worse than method 1, although the presence of the outlying open triangle does suggest this.

The decision surfaces are modified in a way that seems intuitively reasonable (Figure A4.2).

It appears that, to a first approximation, the error rates are varied in the ratio:*

$$\frac{(\text{No. of errors with } T = +1)}{(\text{No. of errors with } T = -1)} = \frac{V}{1 - V} \cdot \frac{1}{R^2} \tag{A4.3}$$

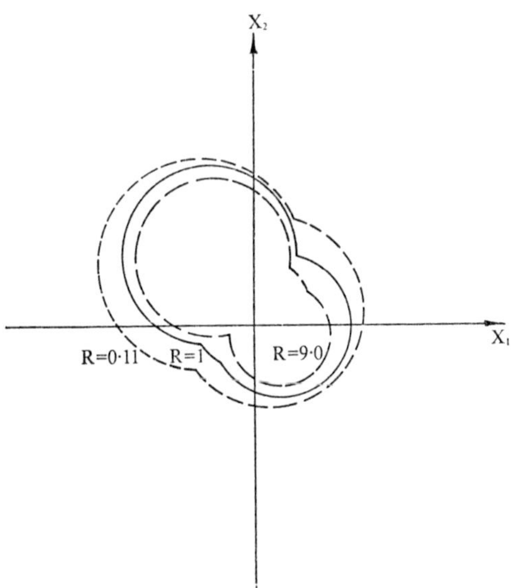

Figure A4.2 Comparison of the decision surfaces obtained using the first method of biasing a classifier to vary the proportions of errors. The data used in this study is described in Appendix 3. (A) This CC encloses the patterns from class 1. (B) This CC encloses the patterns from class 2.

* Other experiments not reported here, have been performed to study the mixing of methods 1 and 2. This conclusion is substantiated by these experiments.

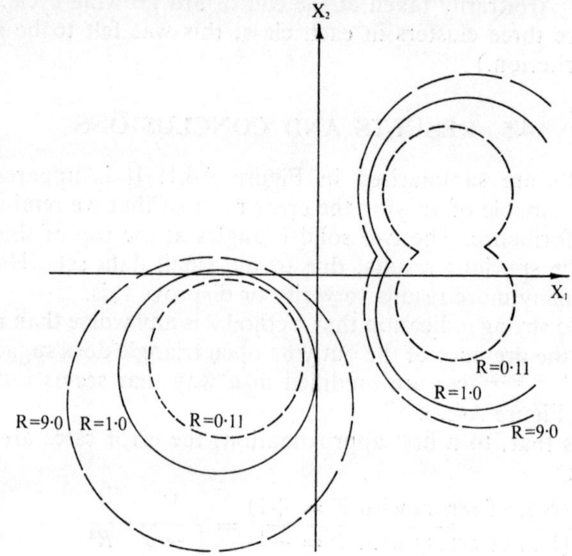

Figure A4.2 (*continued*)

ACKNOWLEDGEMENTS

I should like to thank Professor G. D. Sims for his continued encouragement and for reading the original manuscript. I have enjoyed many hours of valuable discussion with my present and former colleagues: Mr. B. R. Wilkins, Dr. D. W. Thomas, Dr. D. M. Balston and Mr. C. J. W. Mason.

Some of the ideas and results described in this book originated while I was employed by the Electronics Research Laboratory of the Plessey Company Limited. I am grateful for the opportunity to include these.

I am indebted to the following individuals, who have kindly supplied me with data, whose analysis is described in this book. I am also grateful to them for reading the relevant sections.

Dr. C. D. Binnie, Mr. D. S. Lloyd, St. Bartholomew's Hospital, London.
Dr. W. E. Waters, University of Southampton.
Dr. I. Glen, M.R.C., Edinburgh.
Dr. H. Miller, Portsmouth and Isle of Wight Area Pathology Service.
Dr. J. R. Parks, N.P.L., Teddington.

I am pleased to acknowledge the following organisations and individuals who have given me permission to include their copyright material:

Dr. D. M. Balston, Plessey Company Limited.
Dr. D. W. Thomas, University of Southampton.
Mr. D. E. Maynard, London Hospital.
Publishing Manager, British Medical Journal.
Editor, Electro-Psysiological Technolgists' Association.
Editor, Electronics Letters.
Editor, Proceedings of the Institution of Electrical Engineers.
Editor, Conference Publications, Institution of Electrical Engineers.
Commercial Manager, Institute of Physics.
Editor, Information Sciences, American Elsevier Publishing Company Incorporated.

My gratitude is also due to Mrs. J. Y. Wright-Green, Mrs. J. R. Ede & Miss P. Hayes who, among them, typed the manuscript. Finally, I must thank my wife, Eleanor. Her patience and unending encouragement have inspired me beyond measure. To her, my Mother and Father, I dedicate this book.

Subject Index